VISIONS OF NUMBERLAND

A COLOURING JOURNEY THROUGH THE MYSTERIES OF MATHS

ALEX BELLOS
EDMUND HARRISS

BLOOMSBURY
LONDON · OXFORD · NEW YORK · NEW DELHI · SYDNEY

ACKNOWLEDGEMENTS

We'd like to thank Xa Shaw Stewart and Emma Ewbank at Bloomsbury, and Karen Giangreco at The Experiment, for turning our ideas into a glorious book. Thanks also to our agents, Rebecca Carter and Emma Parry at Janklow & Nesbit. We met at the Gathering for Gardner and would like to thank everyone involved for the mathematical inspiration and play, some of which made its way into this book. Edmund's wife Ása and Alex's wife Nat were again hugely nurturing and helpful — we owe it to you both one more time!

Bloomsbury Publishing
An imprint of Bloomsbury Publishing Plc

50 Bedford Square   1385 Broadway
London   New York
WC1B 3DP   NY 10018
UK   USA

www.bloomsbury.com

BLOOMSBURY and the Diana logo are trademarks of Bloomsbury Publishing Plc

First published in Great Britain 2016

Star-Studded Geometry and Diamonds in the Sky © Jay Bonner (www.bonner-design.com). Both designs feature in his forthcoming book *Islamic Geometric Patterns: Their Historical Development and Traditional Methods of Construction*, Springer Verlag | Moroccan Starburst © Marc Pelletier | Scaling Shapes © Chaim Goodman-Strauss | Naked Geometry © James Gyre (www.nakgeo.com) | Hopf Fibration © Henry Segerman | Mirror Balls © Arnaud Chéritat | Taffy puller by Karen Giangreco | Flower photograph © Ásgerður Jóhannesdóttir

British Library Cataloguing-in-Publication Data
A catalogue record for this book is available from the British Library.

ISBN: TPB: 978-1-4088-8898-8

10 9 8 7 6 5 4 3 2 1

Printed and bound in China by C&C Printing.

Bloomsbury Publishing Plc makes every effort to ensure that the papers used in the manufacture of our books are natural, recyclable products made from wood grown in well-managed forests. Our manufacturing processes conform to the environmental regulations of the country of origin.

To find out more about our authors and books visit www.bloomsbury.com. Here you will find extracts, author interviews, details of forthcoming events and the option to sign up for our newsletters.

Sangaku solution: The radius of the larger circle is 10 times that of the smaller one.

INTRODUCTION

Welcome to a magical mystery tour of some of the most fundamental discoveries in maths. Prepare to be surprised, thrilled and awed by images that are as provocative as they are beautiful.

The first section, *Colour*, contains patterns to be coloured in. The second, *Create*, provides guidelines for you to create your own images.

Our journey explores traditional fields like arithmetic and geometry, and also much more modern areas of maths research like graph theory, dynamical systems and algorithms. You'll encounter the two most famous theorems in maths – Pythagoras' theorem and Fermat's last theorem – and also some fascinating problems yet to be solved, such as the Collatz conjecture.

Many of our illustrations derive from the work of geniuses – like Euclid, Isaac Newton, Gottfried Leibniz, Leonhard Euler, Carl Friedrich Gauss, Sophus Lie, Felix Klein and Alan Turing. You don't need to understand any maths to appreciate the images, but we hope they encourage you to contemplate the patterns they reveal.

Look through these pages to see the universe through the eyes of the world's great mathematicians – and then bring it to life with colour.

COLOUR

CURVE OF PURSUIT

The path taken by a point that is always moving towards another moving point

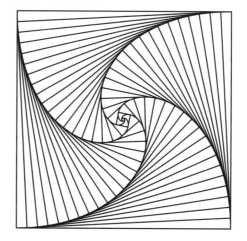

THE CHASE

Here's an appropriate curve to get this book up and running. Imagine four dogs at the four corners of a square, as above. If each dog races towards the dog that is anticlockwise from it, such that both dogs are running at the same speed and at every moment the chasing dog is aiming for the target dog along a straight line, the dogs will follow *curves of pursuit* – in this case, a logarithmic spiral – that meet at the centre of the square. You get the same type of curve if the dogs start from the corners of a triangle or a hexagon. The image opposite is a collage of several of these canine chases.

GEOMETRY

The study of space using points, lines and shapes

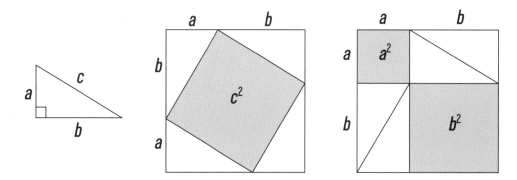

PYTHAGORAS SQUARED

Pythagoras' theorem — the most famous theorem in maths — says that for any right triangle, the square of the hypotenuse (the side opposite the right angle) is equal to the sum of the squares of the other two sides. In other words, for the triangle above, the theorem states that $a^2 + b^2 = c^2$. For a simple geometrical proof of Pythagoras, we can imagine four identical right triangles inside a square with side $a + b$. Arranged one way, the surplus (shaded) area is a square with side c, and arranged another way, the surplus area is two squares with sides a and b. Since the surplus areas must be equal, we have shown that $a^2 + b^2 = c^2$. Ta-da! The image opposite is a grid inspired by this famous proof.

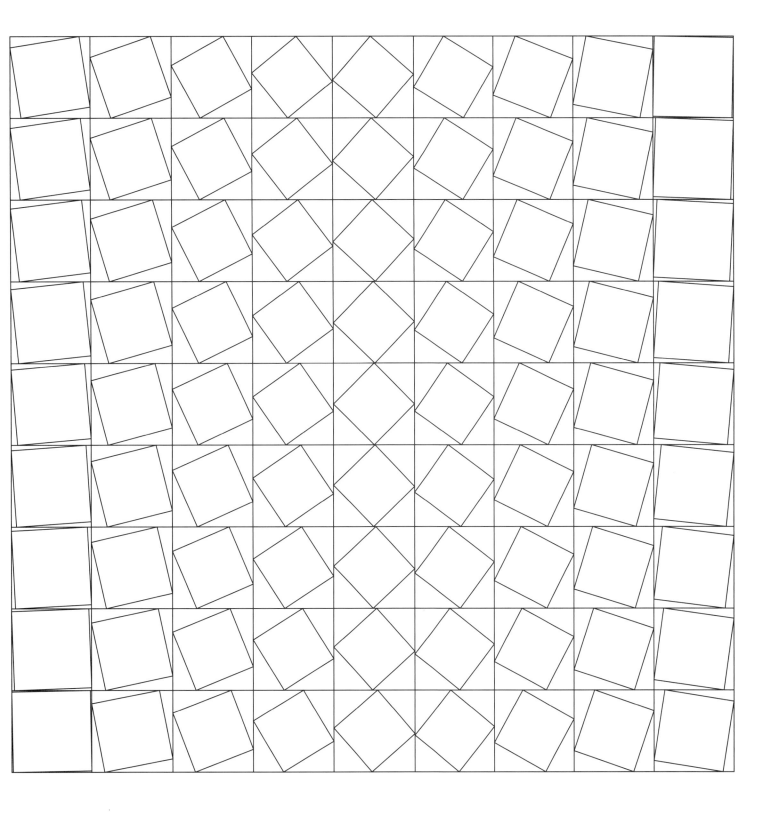

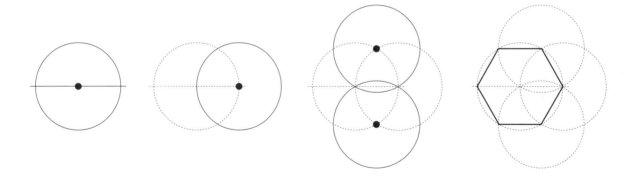

ELEMENTS OF EUCLID

In his book *Elements*, the ancient Greek geometer Euclid set out to construct shapes using only a straightedge, or unmarked ruler, for drawing lines and a compass for drawing circles. The four steps above show how to draw a regular hexagon. The image opposite was also constructed using only a straightedge and compass. We drew the bottom hexagon first, and then used the dotted lines to draw the equilateral triangle to its upper right. With the triangle, the rest of the pattern follows.

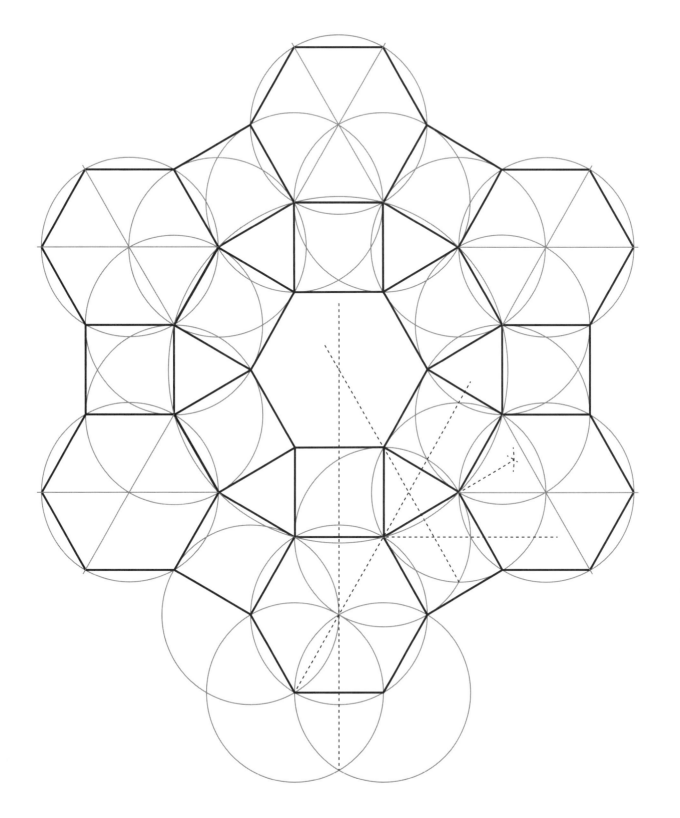

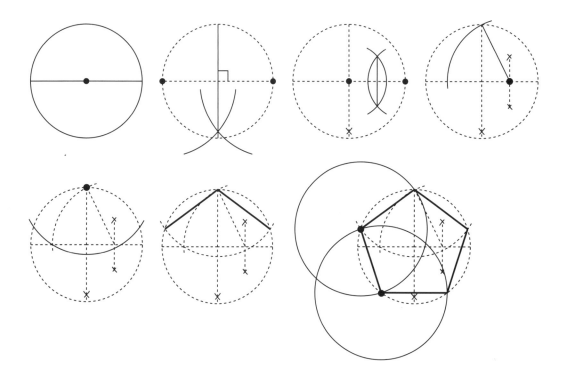

EUCLID'S PENTHOUSE

The seven steps above take you through Euclid's process for drawing a regular pentagon with just a straightedge and a compass. When you arrange regular pentagons together there are always gaps, such as the lozenges in the pattern opposite. (You can avoid gaps when pentagons are not regular — as you'll see later on.)

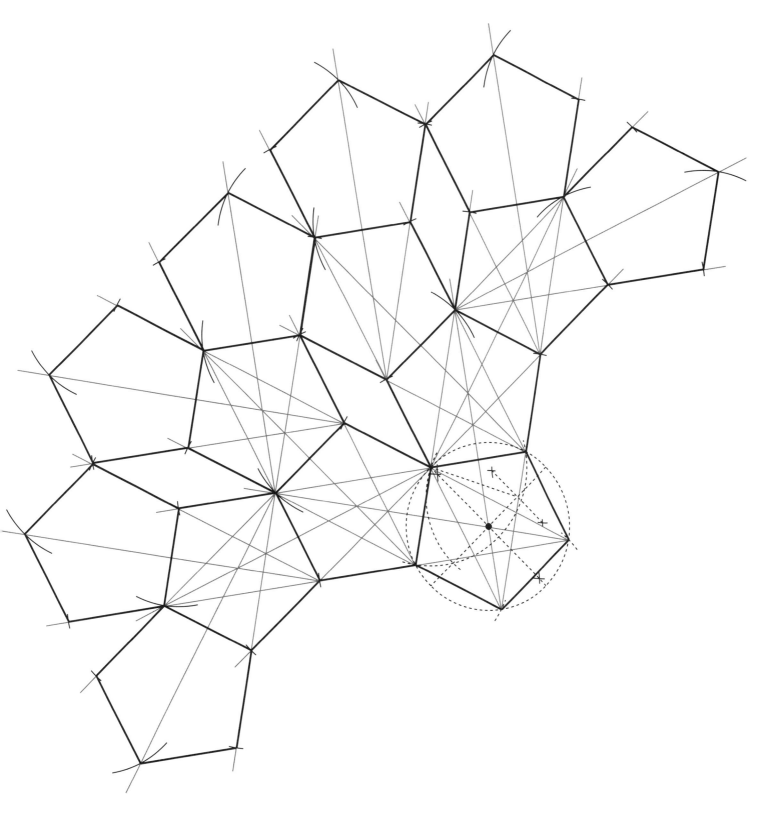

ISLAMIC GEOMETRY

A distinctive design style used by Islamic artists from the ninth century onwards

The repeating patterns are always constructed using only a compass and
a straightedge, a convention the Islamic artists inherited from the Greeks.

STAR-STUDDED GEOMETRY

Islamic geometric design remains a vibrant field, and one of its top specialists is the American artist Jay Bonner, who has
designed patterns for the Grand Mosque and the Kaaba in Mecca. This image of his links 11-, 12-, and 13-pointed stars.

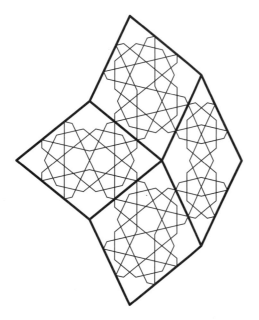

DIAMONDS IN THE SKY

In this image, also by Jay Bonner, a pattern of 14-pointed stars is based on a grid made up from two varieties of rhombus.

MOROCCAN STARBURST

This design, using 16- and 24-pointed stars, is by American geometric artist Marc Pelletier, who was inspired by the patterns he saw on a trip to Morocco.

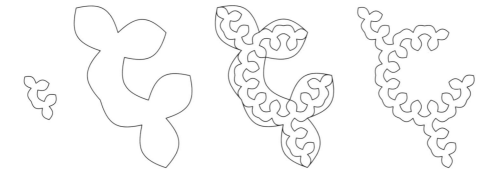

SCALING SHAPES

The aesthetics of Islamic geometric design have inspired many mathematical artists, such as Professor Chaim Goodman-Strauss at the University of Arkansas in the US. This piece, which he calls 'Two plus the square root of three', is constructed like this: (1) Take the small tile above. (2) Enlarge it to $2 + \sqrt{3}$ times its size. (3) Fill the larger tile with smaller ones. (4) Remove the outline of the large tile.

NAKED GEOMETRY

The work of mathematical artist James Gyre, which he calls 'naked geometry', is also heavily inspired by Islamic geometric design.

PERSPECTIVE

Drawing 3D objects on a flat surface as the eye sees them

This particular field of geometry was first developed by artists in the Renaissance.

FIELD OF CUBES

To get the full effect of the 3D illusion opposite, place your eye 10cm directly above the bottom left cube.

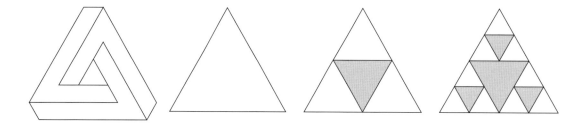

TRIBAR TRIBUTE

The problems of representing three-dimensional objects in two dimensions can lead to fun optical illusions such as the *tribar*, above left. Our brains interpret it as a 3D object, but we quickly see that it could not possibly exist in the real world. The other three images above show the construction of the *Sierpinski triangle*: Divide an equilateral triangle into four smaller triangles and remove the one in the centre. Repeat with the remaining triangles as many times as you want. The image opposite combines both ideas.

TRIBAR CUBED

Here's another impossible object based on the tribar and the Sierpinski triangle.

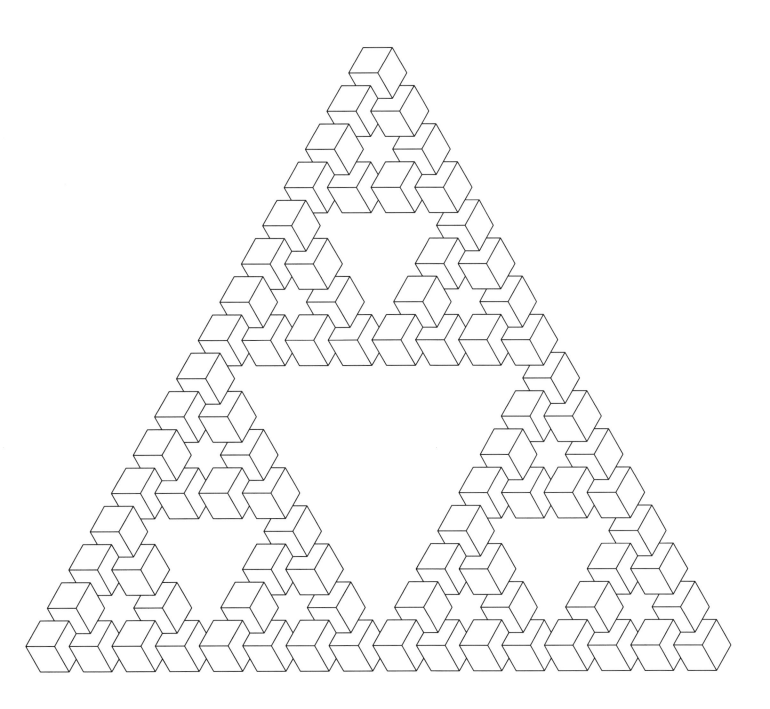

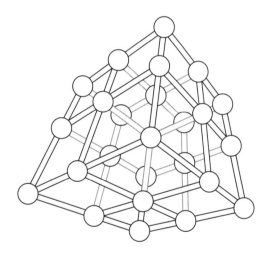

PING PONG PANDEMONIUM

Imagine stacking ping pong balls in rows and columns, as in the block floating above. Now extend the block by adding balls upwards, lengthways and widthways, so it becomes a much bigger stack of balls suspended in the air. Erase all the connecting spokes. The image opposite is a view of this mega-stack from an interesting angle. The effect of some balls appearing in lines and others not is similar to what you see when you drive past an orchard of trees planted in rows. Depending on your position, you will see corridors through the trees in some directions, and not in others.

4D PING PONG

Even though we live in three-dimensional space, maths allows us to describe four – or more – dimensions. We cannot 'see' an object in 4D space, but we can see its 'shadow' in two dimensions. For example, if we stack ping pong balls in four dimensions, the image opposite is their 'shadow' in two.

MENGER SPONGE

A cube with progressively smaller cubes cut out of it

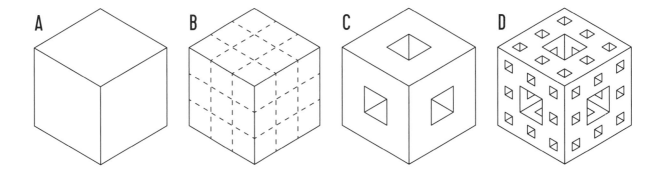

CUBES ALL THE WAY DOWN

First described in 1926 by the Austrian mathematician Karl Menger, the *Menger sponge* is constructed like this: If we think of the original cube (A) as being made out of 27 smaller cubes (B), the first step is to cut out the cube at the very centre, plus the six outer cubes that share a face with it (C). The second step is to divide each of the 20 remaining cubes into 27 even smaller cubes, and cut out the same elements (D). Repeat this once more to get the image opposite.

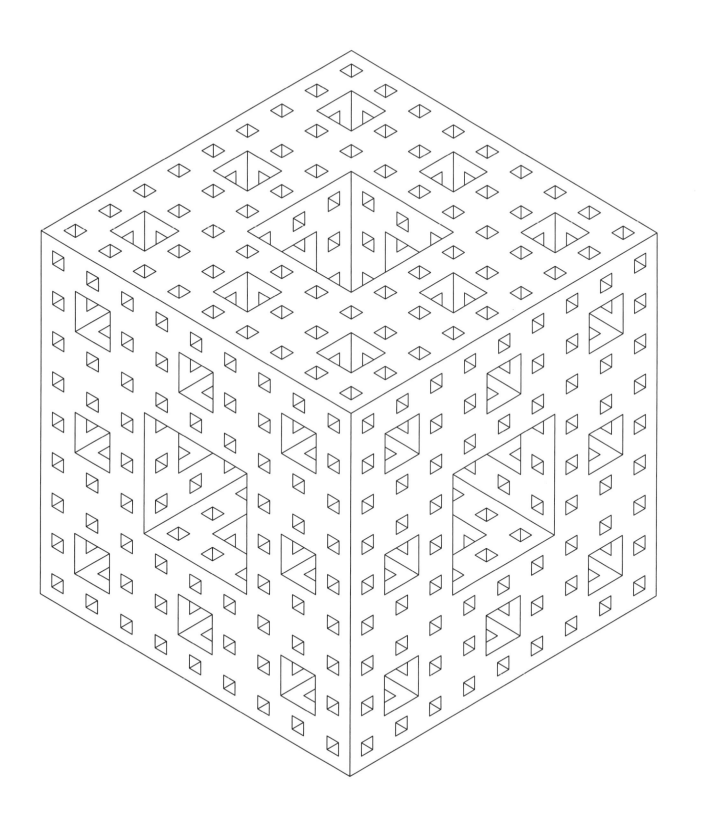

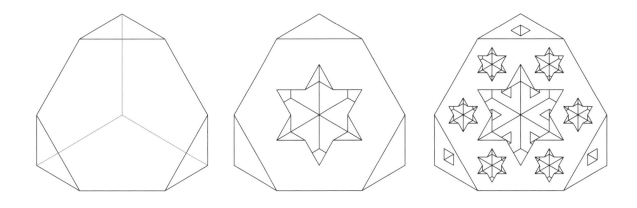

MENGER SLICE

If you cut through a cube at a certain angle, you can create a hexagonal face (as above left). When you cut a Menger sponge the same way, the cross-section is a constellation of six-pointed stars.

TOPOLOGY

The field concerning the properties of space that are not changed by stretching, bending or squeezing, but by tearing, perforating and gluing

HAIRY BALL

The so-called *hairy ball theorem* states that it is impossible to comb a hairy sphere flat without there being at least one tuft left sticking up. A consequence of this theorem is that there is always at least one point on the Earth's surface where there is no wind. If you stretch, bend or squeeze a hairy ball into a new shape, this new shape will also be impossible to comb flat.

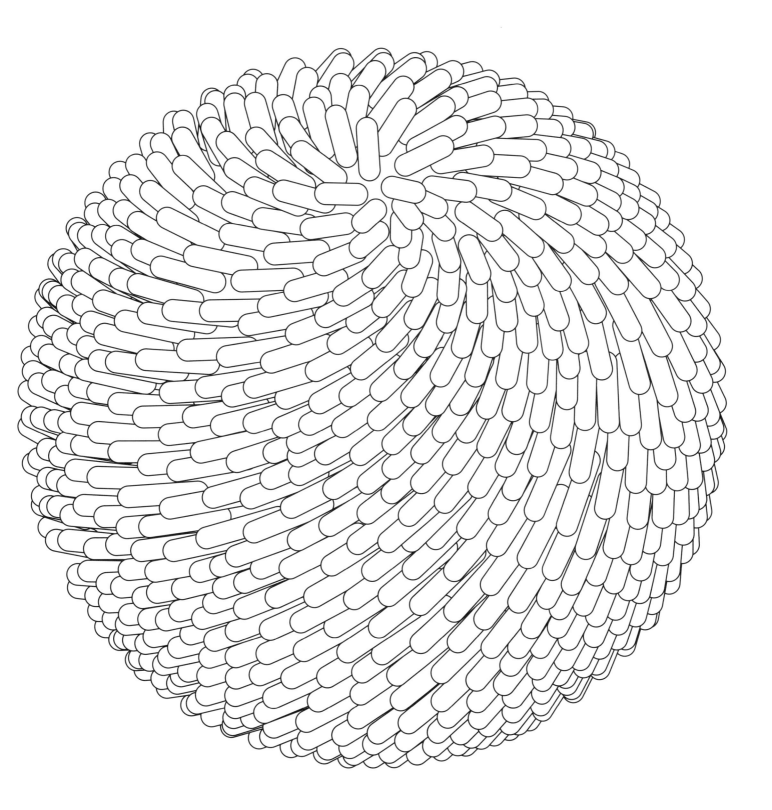

DONUT WRAP

There is no 'hairy donut theorem', but if there was, it would state that it *is* possible to comb a hairy *torus* – that's the maths term for a donut shape – so that all the hair lies flat. The image opposite shows a single ribbon wrapped around an invisible torus. The ribbon lies flat on the torus and joins up with itself. If the torus was hairy, you would be able to comb it flat by following the contours of the ribbon.

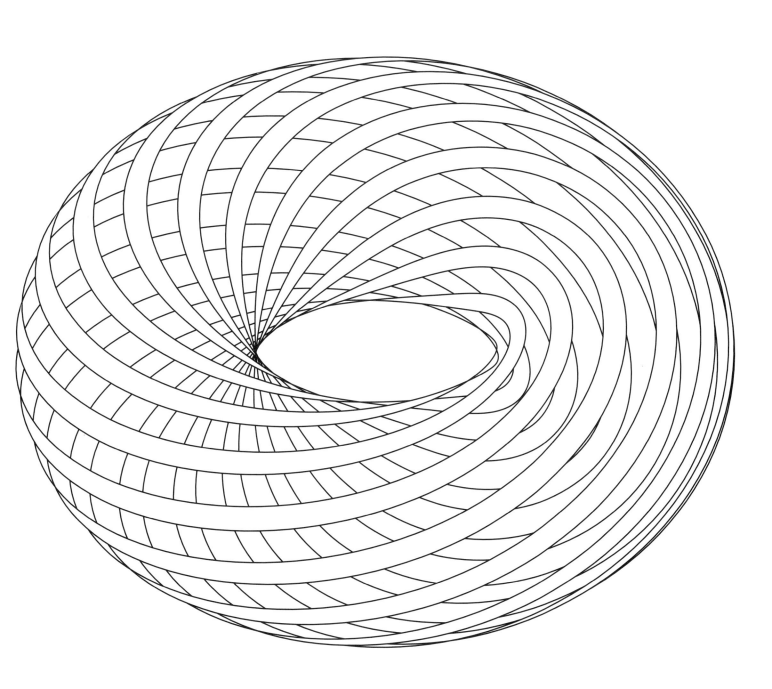

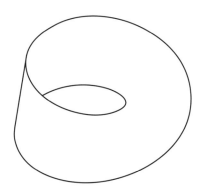

MÖBIUS BEADS

A Möbius strip is a band of flat material that only has one side, like the one above. You can make one by taking a long strip of paper, making a single twist, and then joining the ends. (To prove to yourself that there is only one side, try drawing a line down the centre of your paper Möbius strip.) Opposite is a Möbius strip made from four separate chains of beads. Follow the beads with your finger to see just how they twist around.

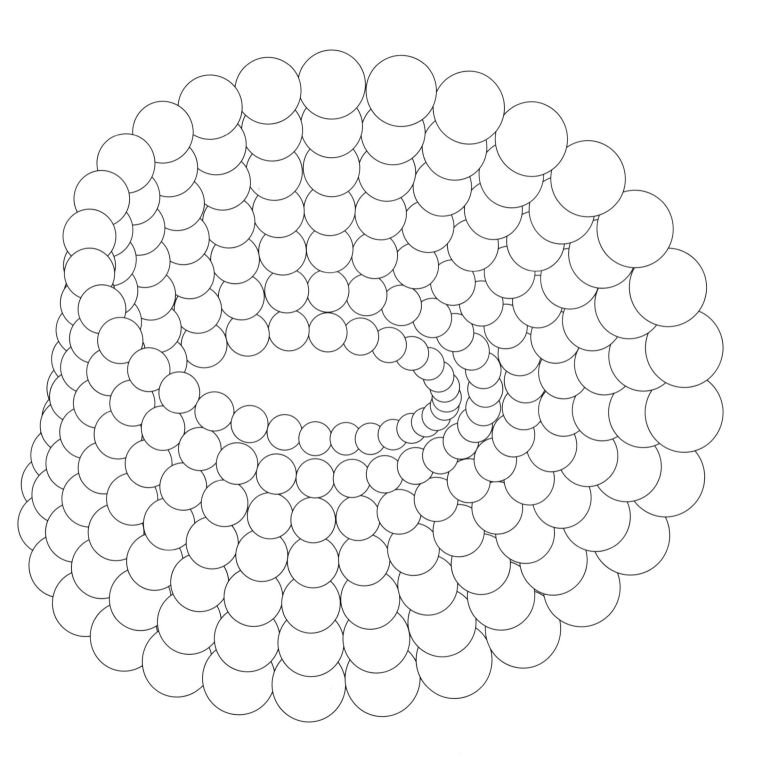

HOPF FIBRATION

Usually when we talk about a sphere we mean a sphere in three dimensions, which is defined as the set of points equidistant from a fixed point (at the sphere's centre). We can use the same definition to describe a 'sphere' in four dimensions – although we cannot visualise it, since the real world has only three spatial dimensions. However, in 1931 German mathematician Heinz Hopf discovered a way to visualise the surface of the 4D 'sphere' in three dimensions, as circles threaded around each other like fibres. Opposite, mathematician Henry Segerman has found a stunning way to turn Hopf's 3D object into a 2D image.

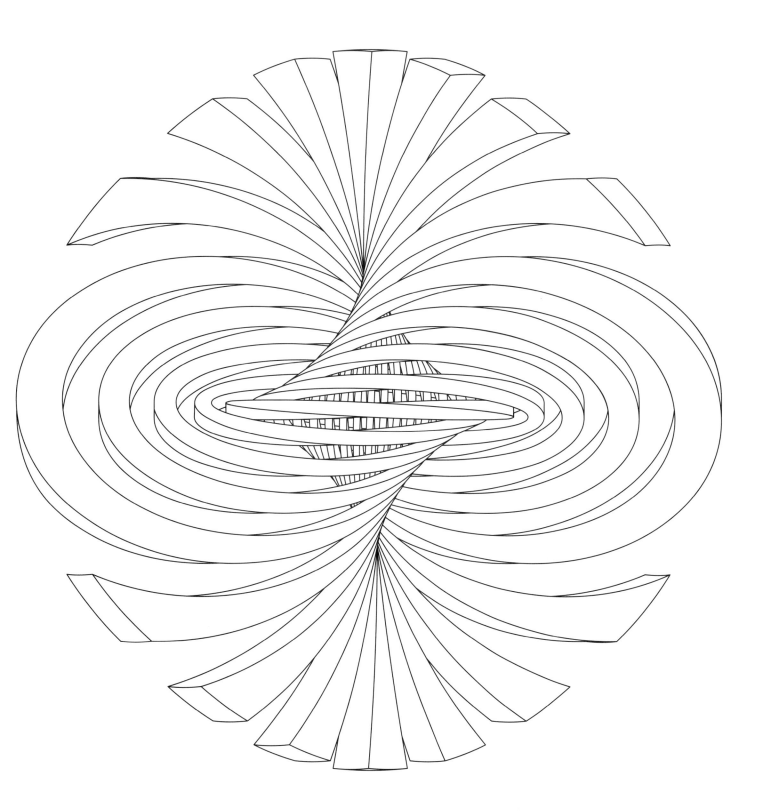

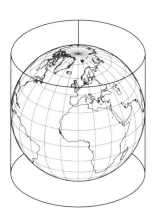 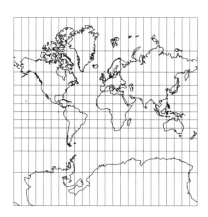 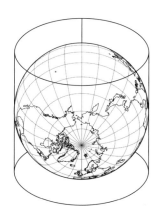

FUNNY OLD WORLD

The Mercator projection of the world – the standard map used by sailors, above – takes the globe and projects it onto a cylinder with its axis parallel to the Earth's. The image opposite shows what happens if the globe is projected onto a cylinder with its axis *perpendicular* to the Earth's. In a sense, the distortions opposite are no more 'wrong' than those in the map we are used to.

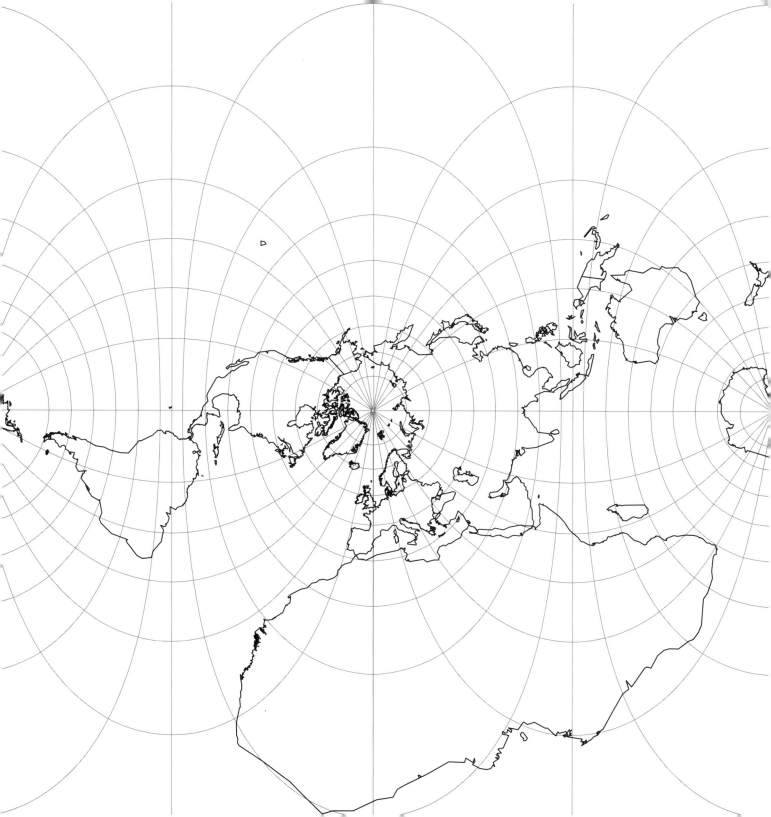

NUMBERS

Looking at arithmetical patterns by representing numbers as dots

PRIME LINES

Here are the numbers from 1 to 36 in order, represented as groups of dots. Where possible, the dots are arranged in a square or a rectangle, which gives us a visual sense of each number's arithmetical properties. A square with a side length of y dots is the number y^2, and a rectangle with side lengths of y and z dots is the number $y \times z$. If a number cannot be arranged as a rectangle or square, and is therefore a long line of dots, it is a *prime number*, the name given to those numbers only divisible by themselves and 1. Note that, after the first row, the prime numbers are always in either the first or fifth column.

HEAVEN'S ELEVEN

The previous image showed how to break down the numbers from 1 to 36 into multiplications. This image, a *Ferrers diagram*, shows how to break down a single number into additions. It shows all the ways of separating 11 dots into groups (represented here as rows). For example, starting from bottom left and moving upwards, we have a single row of 11 dots, then a row of 10 dots and a row of 1 dot, then a row of 9 dots and a row of 2 dots. When numbers are broken down in this way, each arrangement is called a *partition*. There are 56 partitions of 11.

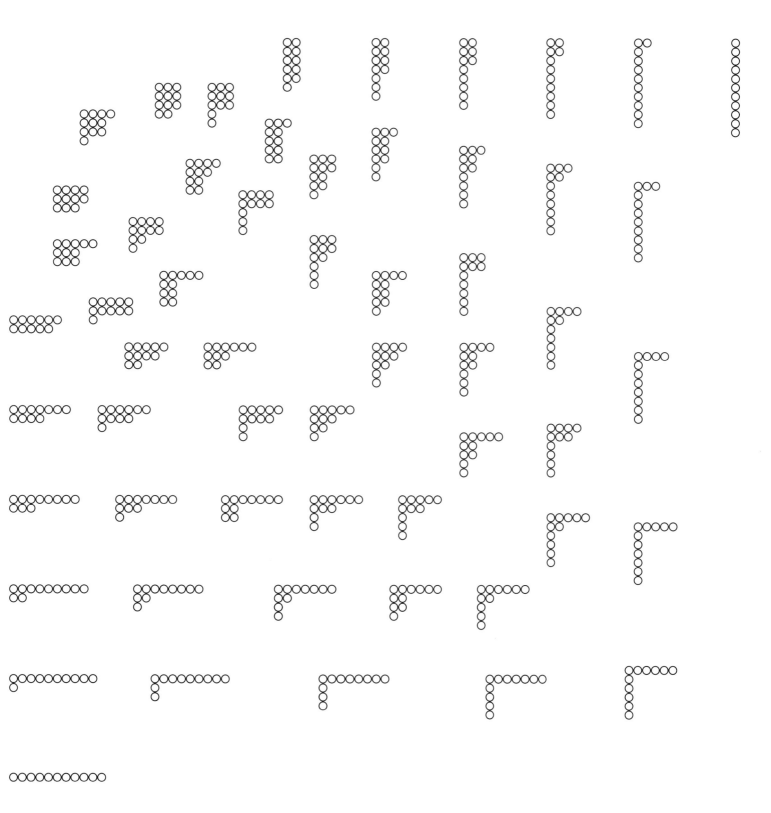

SEQUENCES

A sequence is an ordered list of numbers

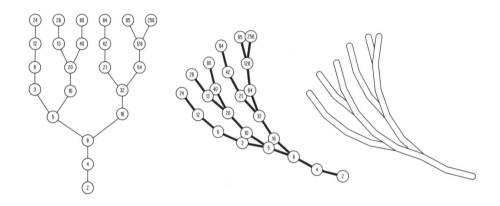

THE COLLATZ CONJECTURE

In 1937 the German mathematician Lothar Collatz invented the following rule: Pick a number. If it is even, divide it by two. If it is odd, triple it and add one. Continue for as long as you can. Let's try it with 13. It is odd, so we triple and add 1 to get 40. Forty is even, so we halve it to get 20. The full sequence goes: 13 → 40 → 20 → 10 → 5 → 16 → 8 → 4 → 2 → 1. Collatz thought all numbers should lead to 1, but he couldn't prove it. His conjecture is now a famous unsolved problem – it's easy to describe, but mathematicians are nowhere close to a solution.

The first image above illustrates the Collatz sequences as a tree going from the top to the bottom. (It condenses the rule for odd numbers into 'triple, add one *and* divide by two' – so we go straight from 13 to 20.) Now consider the centre image above, starting at the bottom with 2. Whenever the number doubles, the branch bends slightly clockwise. Otherwise, it bends slightly anticlockwise. Smoothed out, the branches look very like tentacles.

The image opposite includes all the sequences starting with numbers below 10,000. (The 'suckers' on the tentacles show the start of branches that head off *beyond* 10,000.) The wild, organic-looking structure reveals how a simple, orderly rule can create havoc.

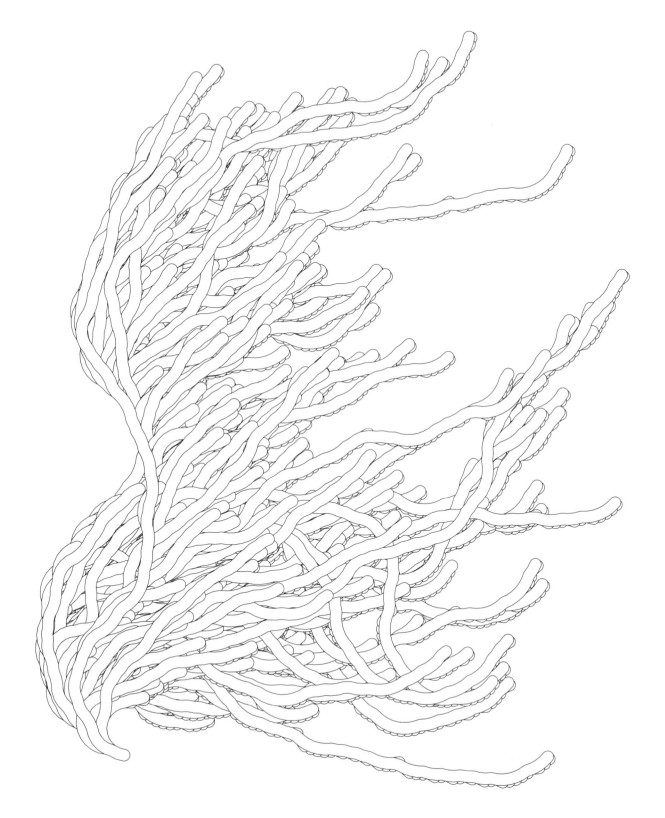

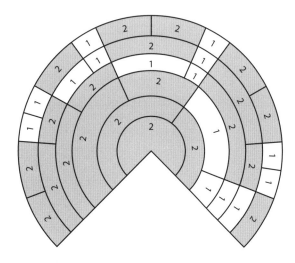

THE KOLAKOSKI SEQUENCE

The Kolakoski sequence contains only 1s and 2s, which appear in either a run of one or a run of two, as shown below.

$$\underline{1}, \underline{2, 2}, \underline{1, 1}, \underline{2}, \underline{1}, \underline{2, 2}, \underline{1}, \underline{2, 2}, \underline{1, 1}, \underline{2}, \underline{1, 1} \ldots$$

1 2 2 1 1 2 1 2 2 1 2

It has the mathematically interesting property that the lengths of the runs of 1 and 2 recreate the original sequence. The Kolakoski sequence (with or without the initial 1) is the only sequence with this property. The rings above and opposite illustrate the sequence without the initial 1. The runs of 1 and 2 in each ring are described in the next smaller ring.

The mathematician William Kolakoski first discussed this idea in 1965. Whether the ratio of 1s to 2s approaches 50/50 as you add more and more terms of sequence remains an open question.

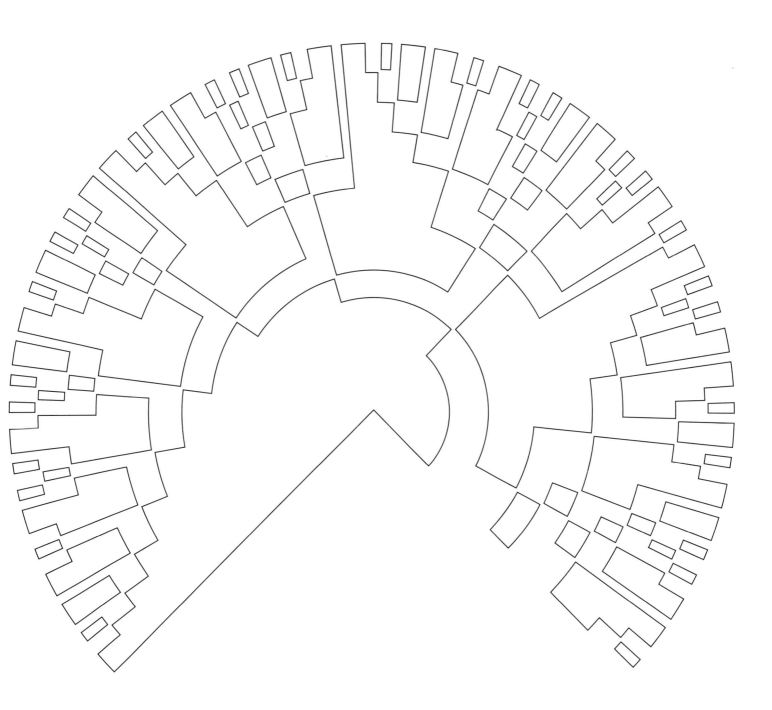

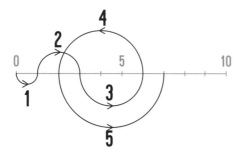

THE RECAMÁN SEQUENCE

Named after the Colombian maths educator Bernardo Recamán Santos, the *Recamán sequence* is a 'jumping' sequence in which the length of the jump increases by 1 at every stage. The rule is that you must always jump backwards if you can, but if jumping backwards would take you to a negative number or a number already in the sequence, then you must jump forwards. Above are the lengths and directions of the first five jumps, represented as semicircles between markers on the number line (in grey). You start at 0 and jump forwards to 1. The sequence continues: 0, 1, 3, 6, 2, 7, 13, 20, 12, 21, 11, 22 . . .

Opposite, the invisible number line is the diagonal starting from the bottom left. Mathematicians find it fascinating that this simple rule generates such a haphazard sequence, which seems to sit on the border between order and chaos.

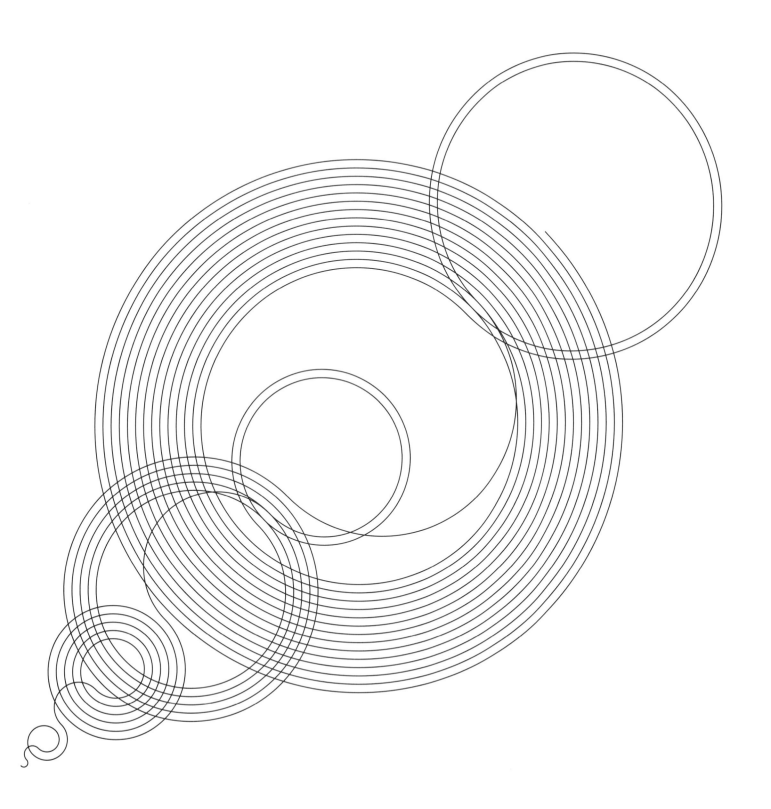

THE THUE-MORSE SEQUENCE

Axel Thue and Marston Morse, a Norwegian and an American, first studied this sequence a century ago. It can be generated using only As and Bs. To make the 'opposite' of any pattern of As and Bs, we will replace all the As with Bs and vice versa. To generate the sequence, we start with AB. Then we create the opposite pattern, BA, and add it to the end to get ABBA. Then we create the opposite of ABBA, which is BAAB, and add *that* to the end to get ABBABAAB. Then we add the opposite of ABBABAAB, and so on. Drawing a line between the As and Bs, as above, produces a sawtooth curve, which is extended over rows and columns in the image opposite.

The Thue-Morse sequence appears in many areas of maths, including algebra, number theory and computer science. It has many interesting properties, such as the fact that no pattern of As and Bs will ever repeat three times in a row. The Dutch grandmaster Max Euwe impressed his peers by using the Thue-Morse sequence to model a chess game where A and B are different moves. At the time, the rules of chess stated that if the same sequence of moves happens three times in a row, the game is declared a draw. Euwe had therefore proved that an infinitely long game of chess was theoretically possible.

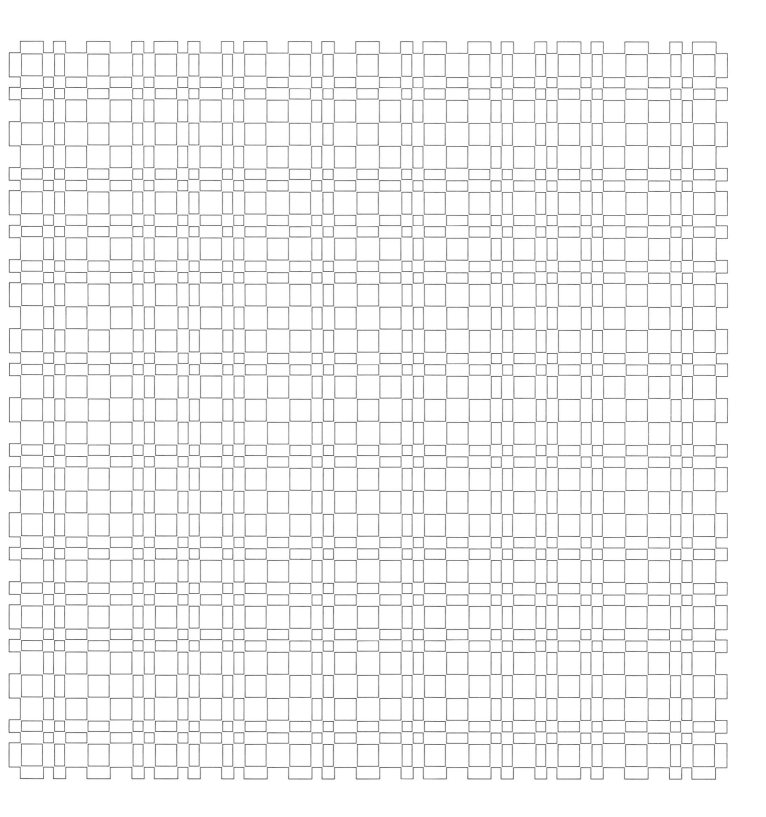

COMPLEX NUMBERS

Numbers that include a multiple of the square root of −1

In the sixteenth century, mathematicians came upon the idea of the square root of minus one. It confused them and many rejected the idea as absurd. How could minus one possibly have a square root, they asked, since whenever you square a number the answer is positive? However, mathematicians soon realised that if they treated the square root of minus one just like any other number (adding, subtracting, multiplying and dividing it) it was useful for solving certain equations. This led over time to its acceptance — and to a new understanding of what numbers are.

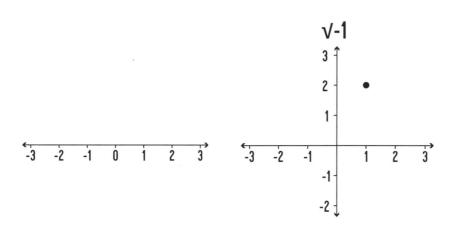

GAUSSIAN PRIMES

The *number line*, above left, is a way to visualise numbers as points on a horizontal line that goes from minus infinity on the left to infinity on the right. The *complex plane*, above right, is a way to visualise complex numbers as points on a plane: the horizontal axis is the number line and the vertical axis represents multiples of the square root of minus one. (For example, the dot is 1 unit along horizontally and 2 units up vertically — so it represents the complex number '1 plus 2 times the square root of minus one', written $1 + 2\sqrt{-1}$.) The *Gaussian primes*, illustrated right, are the complex number equivalent of prime numbers.

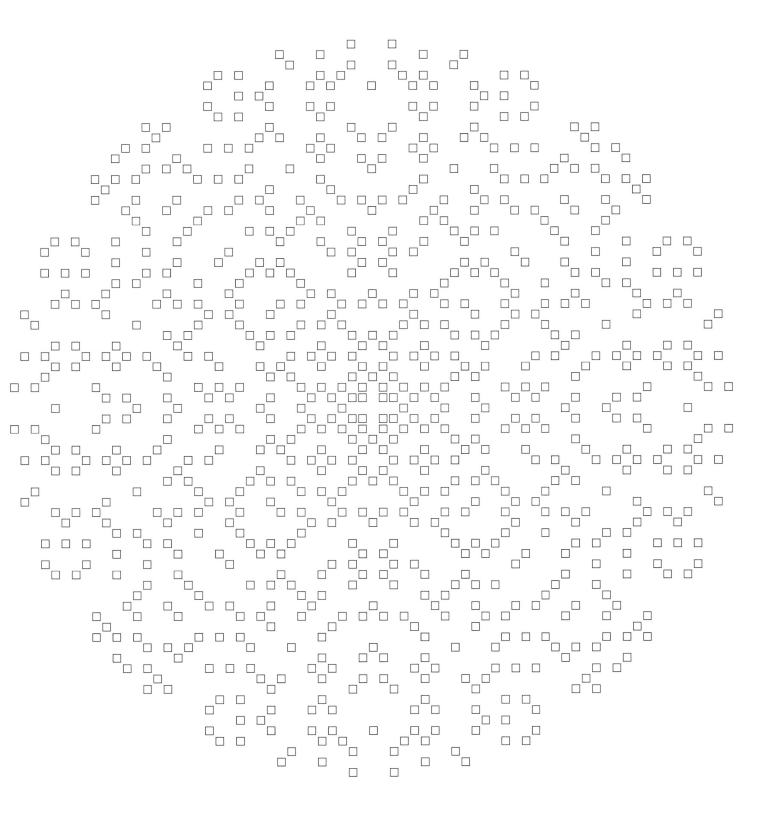

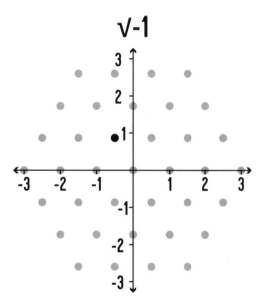

√-1

EISENSTEIN PRIMES

The *Eisenstein integers* (in grey above) are the complex numbers that include a multiple of the complex *cube* root of 1, which is $-\frac{1}{2} + \frac{\sqrt{3}}{2}\sqrt{-1}$, or the bold dot above. The equivalent of prime numbers for the Eisenstein integers are the *Eisenstein primes*. Their positions are the tiny hexagons on the complex plane opposite.

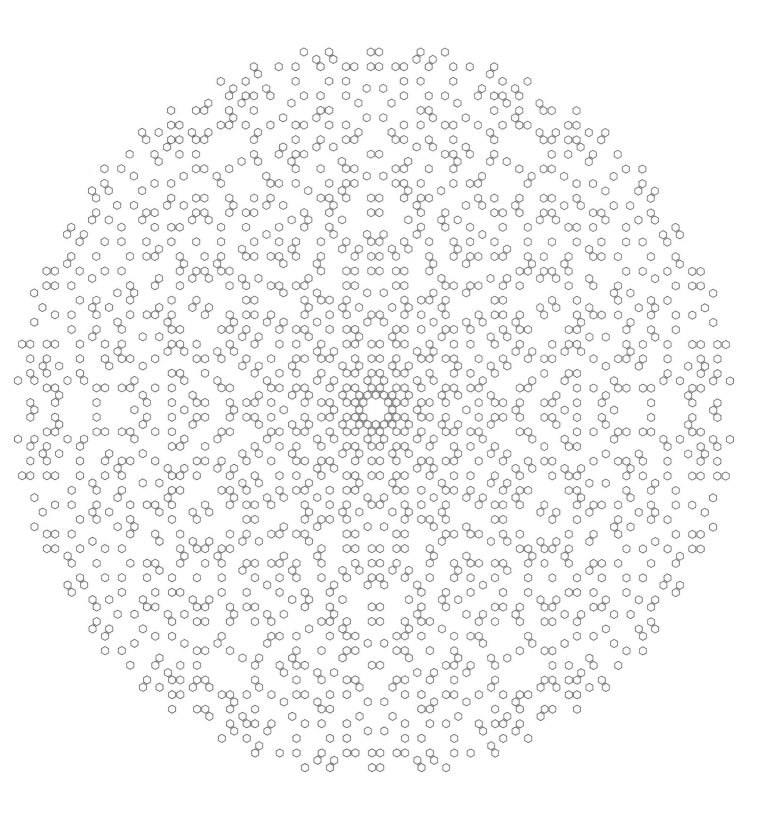

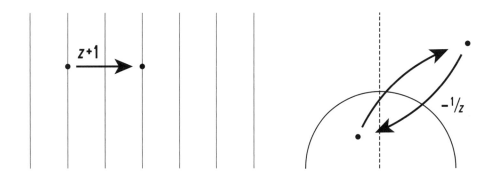

MODULAR FORMS

The image opposite is a combination of the two 'mappings' above on the upper half of the complex plane. In the left one, any point z moves to $z + 1$, and in the right one, any point z moves to $-1/z$, which has the effect of taking a point from inside the semicircle to outside and vice versa. The symmetries in the image opposite are the same symmetries in a mathematical object called a *modular form*, which was crucial to Andrew Wiles' proof of Fermat's last theorem.

SPIDER EGGS

If the mappings of the previous page are applied using different mathematical parameters – for those who really want to know, the ring of Gaussian integers – you get the image opposite, which is called the *Schmidt arrangement* of the Gaussian integers. Not one for arachnophobes! This type of image was popularised by Katherine E. Stange at the University of Colorado Boulder in the US.

KALEIDOSCOPE

Here is another beautiful Schmidt arrangement – this time, of the Eisenstein integers.

REFLECTIONS

Modelling the passage of light through spherical mirrors, which requires complex numbers

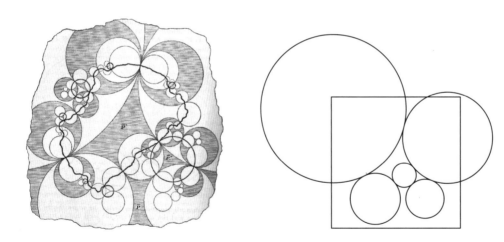

MIRROR BALLS

The diagram above, from an 1897 book by the German mathematician Felix Klein, is one of the great images of mathematics. It shows the patterns of reflection of five mirrored spheres of different sizes. The image — recreated opposite by the French mathematician Arnaud Chéritat — is historically significant because it marked the emergence of a new type of geometry that explores reflections in circles and spheres. Drawn in an age well before computer-generated images, it also demonstrates a stunning level of geometric insight and creativity.

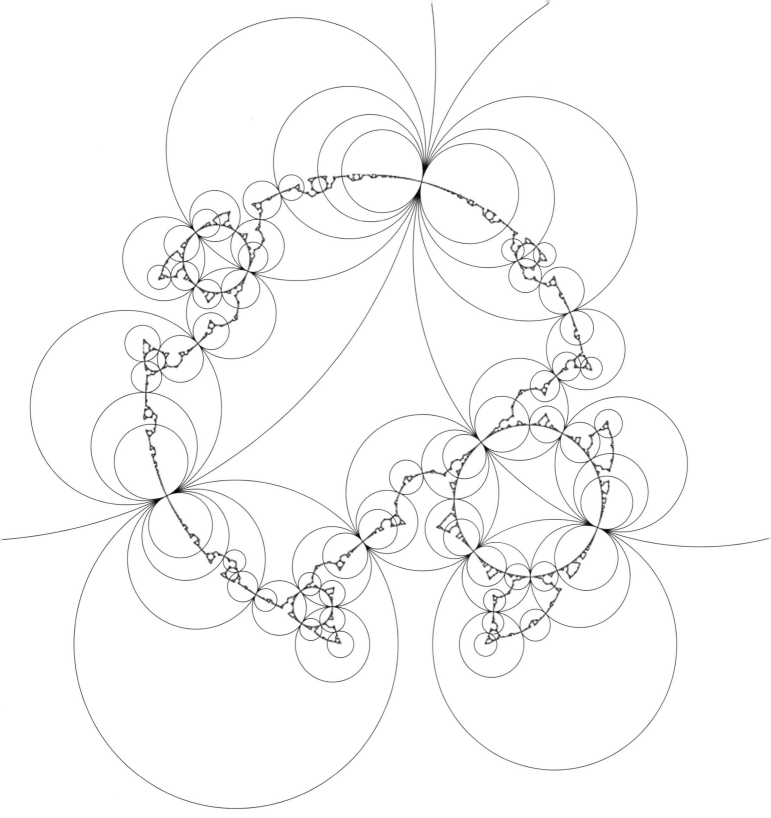

FRIEZE PATTERNS

Geometric patterns that repeat horizontally

In art, a *frieze* is a horizontal band of decoration on a wall. In maths, a *frieze pattern* not only repeats horizontally, but may have one of seven different types of symmetry, which are known as the seven *frieze groups*.

MING FRIEZES

These seven patterns come from porcelain vases made during China's Ming Dynasty (1368–1644). There is one example from each of the seven frieze groups.

ICELANDIC FRIEZES

These seven patterns were used in Iceland for knitting and embroidery between the sixteenth and eighteenth centuries. Each example displays a symmetry from one of the seven different frieze (or should that be freeze?) groups, in the same order as the Ming examples on the previous page.

ROTATIONS

Spinning an object around an axis

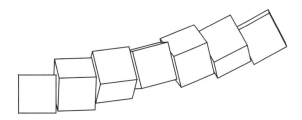

CUBE POMPOM

Start with a single cube. Imagine it is the centre of a pompom, with rigid strands radiating outwards in all directions. Along every strand there are six more cubes, each twisted 30 degrees around the strand relative to the previous cube. One strand is shown above. (The leftmost cube is the central one.) This image is inspired by an idea of Greg Egan, an Australian computer scientist and science fiction writer.

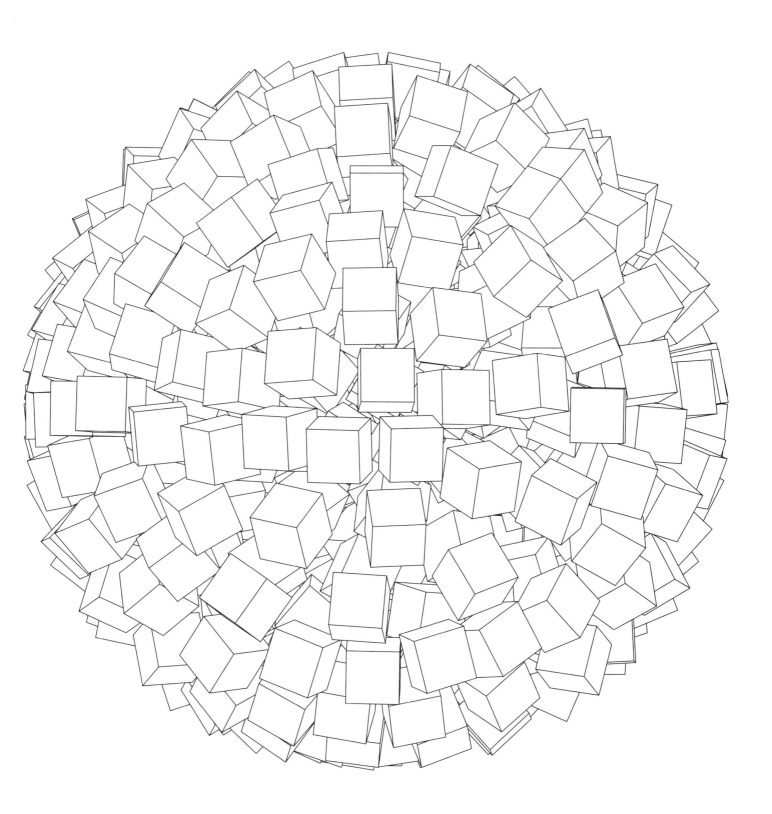

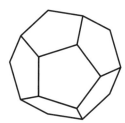

DODECA-POMPOM

This pompom is built on the same idea as the previous image, but with *dodecahedrons*, which have 12 faces — each one a regular pentagon.

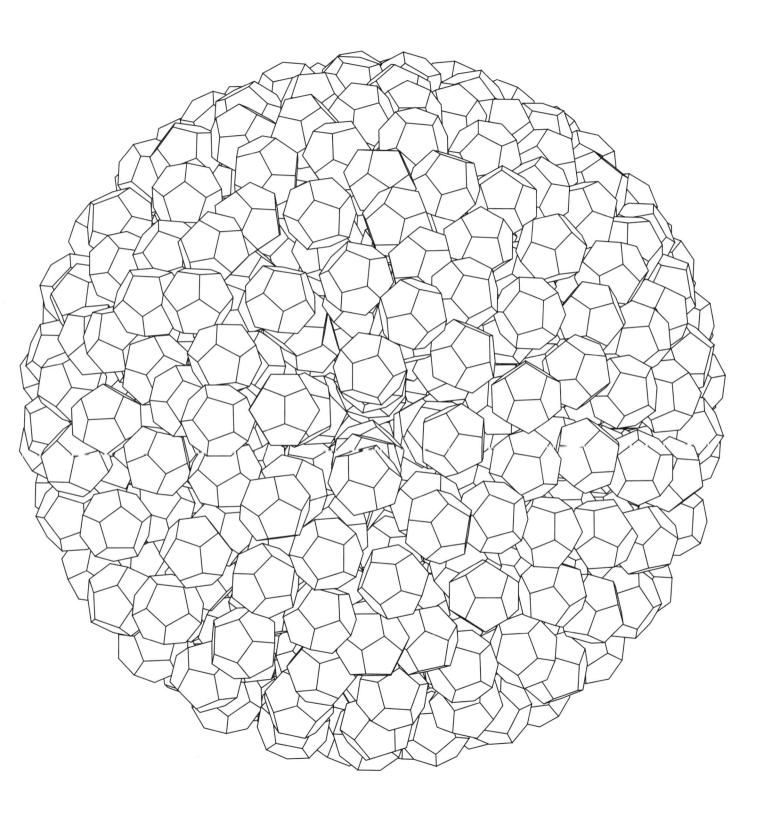

LIE GROUP

A tool for studying objects with certain symmetries

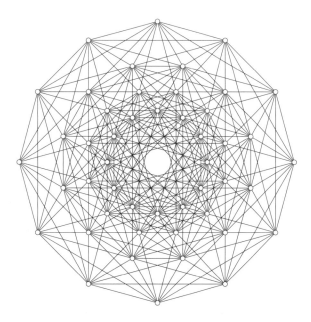

THE ONLY WAY IS E$_6$

Lie groups are a tool used to study objects whose symmetries can be understood through smooth motions. For example, a sphere can spin any amount, in any direction, while appearing the same. (A cube, on the other hand, can be rotated exactly 90 degrees to a new position that looks identical, but it will not look the same *throughout* the rotation.) Above is the most-studied Lie group, called E$_8$, which has 248 dimensions. The group E$_6$ has 78 dimensions and part of its structure is illustrated opposite.

This connection between symmetry and motion has meant that Lie groups – named after the nineteenth-century Norwegian mathematician Sophus Lie – are of fundamental importance for theoretical physicists researching the behaviour of subatomic particles.

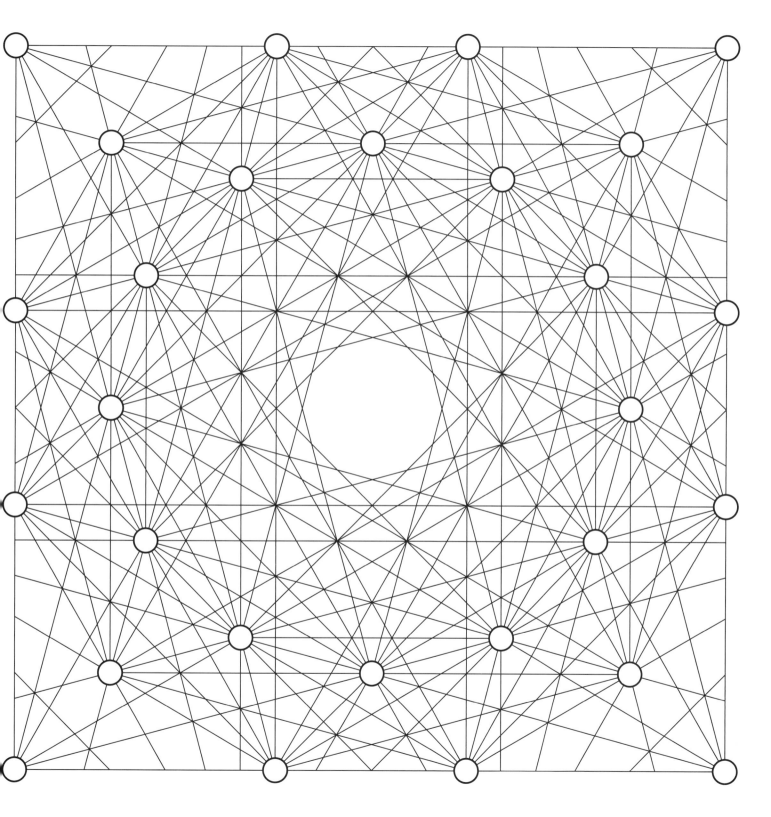

TILING

If many copies of a shape (or set of shapes) can cover a flat surface with no gaps or overlaps, then it is said to 'tile the plane'

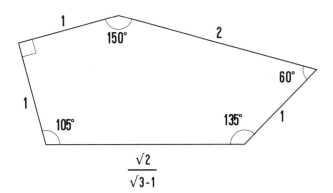

THE FIFTEENTH PENTAGON

In 2015 Casey Mann, Jennifer McLoud and David Von Derau of the University of Washington Bothell in the US made international news when they discovered a new pentagon (above) that can tile the plane. It is the first type of pentagon to be discovered in more than three decades and only the fifteenth since the German mathematician Karl Reinhardt began the classification of pentagon types in 1918.

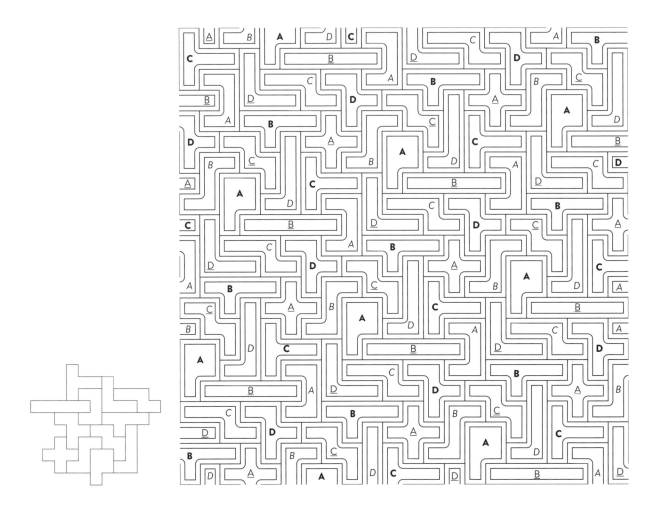

PAINTING PENTOMINOES

Here we've combined tiling, pentominoes and colouring. A *pentomino* is like a domino, but made of five adjacent squares instead of two. The 12 different shapes of pentomino can be combined to make a larger shape (above) that tiles the plane. The *four-colour theorem* states that any tiling of the plane requires four colours — at most — so that no two adjacent tiles share the same colour. In fact, some patterns — like this one — require only three. Here you can try out both! Opposite, we have given each tile an inside and a border. Pick four colours for A, B, C and D, and follow the key above to colour *inside* the tiles. Then, pick three more colours — one each for <u>underlined</u>, **bold** and *italic* letters — and follow the key to colour the borders. Thanks to Alexandre Muñiz for coming up with the idea for this image.

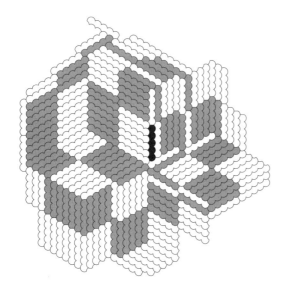

FENCE ME IN

The tile used in this image is a record-breaker! Here's why: Consider the one tile in black above. It is surrounded by a 'fence' of nine identical tiles in white that touch its boundary at every point. The white fence is surrounded in the same way by a fence of 27 tiles in light grey, which is surrounded by a third fence in white, a fourth in grey and a fifth in white. It is impossible to build a sixth fence without leaving a gap somewhere. Therefore, we say this tile's *Heesch number* (named after the German geometer Heinrich Heesch) is 5. Some shapes, like the square, have an infinite Heesch number. But no other known tile has a higher *finite* Heesch number than this one, which was designed by Casey Mann.

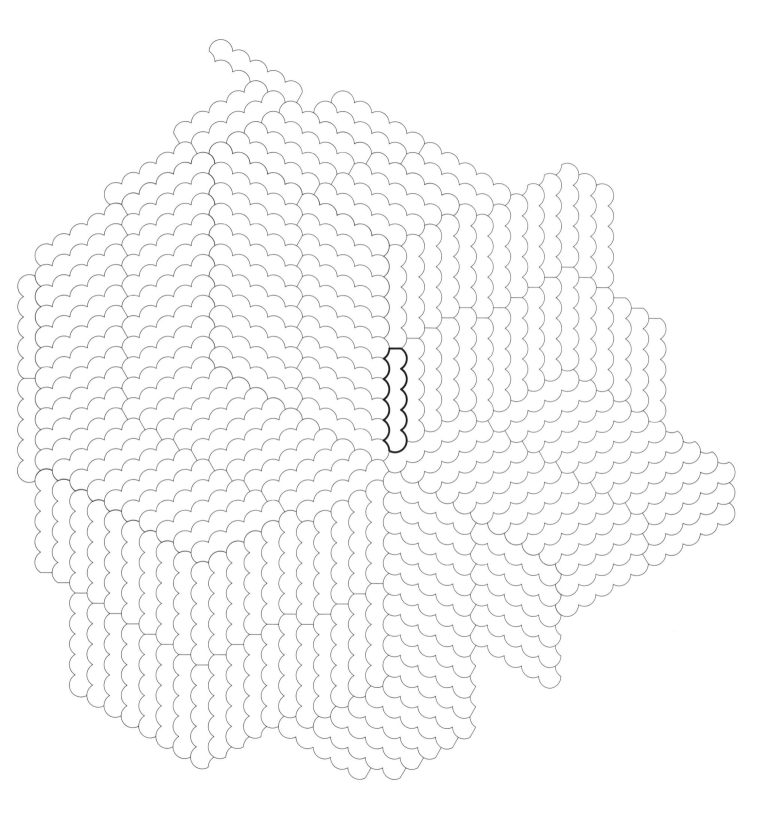

LAND OF THE RISING SUMS

Maths puzzles from Japan

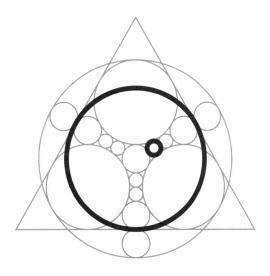

SANGAKU

A fascinating episode in maths history took place in Japan between the seventeenth and nineteenth centuries, when it became common to draw mathematical diagrams on wooden tablets and hang them outside temples. These *sangaku* were pieces of geometry – such as a puzzle or a proof – and were distinctive not just for their mathematical content but also for their beauty. This sangaku is a puzzle: Can you work out how many times bigger the radius of the large bold circle is than the radius of the small bold circle? It was hung at a temple near Nagoya in 1865, and is credited to a 15-year-old boy, Tanabe Shigetoshi. *Answer on the copyright page.*

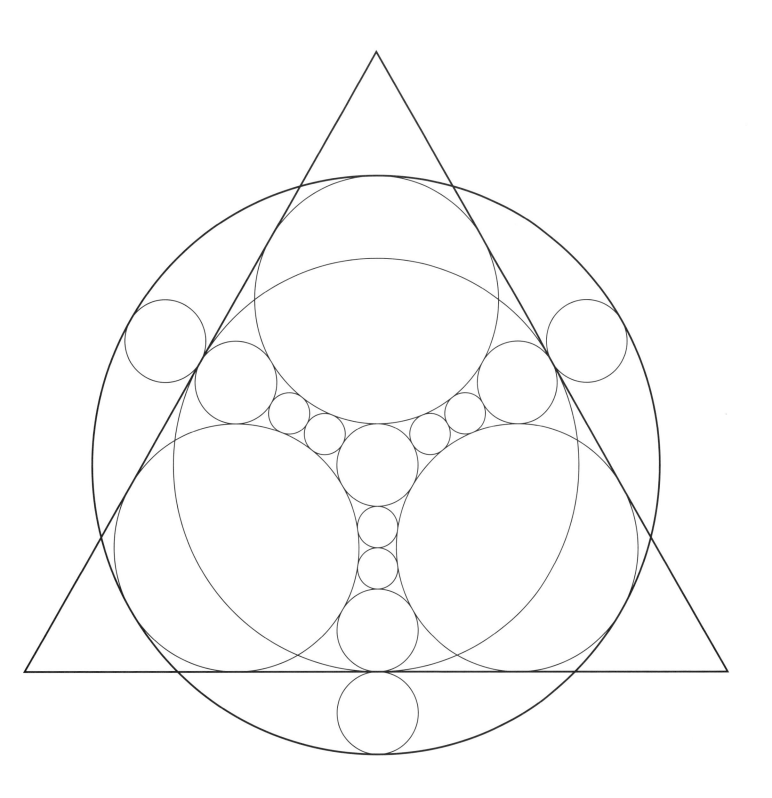

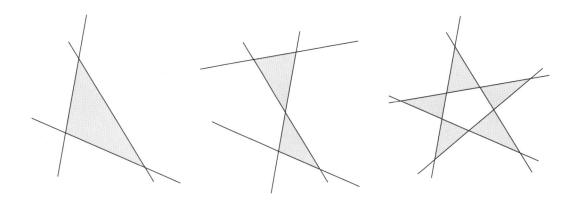

KOBON TRIANGLE

In the mid-twentieth century Kobon Fujimura was a well-known author of Japanese puzzle books. He also asked the following question, which has led to serious geometrical research: Given a certain number of lines, what is the maximum number of non-overlapping triangles you can make? With three, four and five lines, the answers (illustrated above) are 1, 2 and 5. With nine lines, opposite, the maximum is 21.

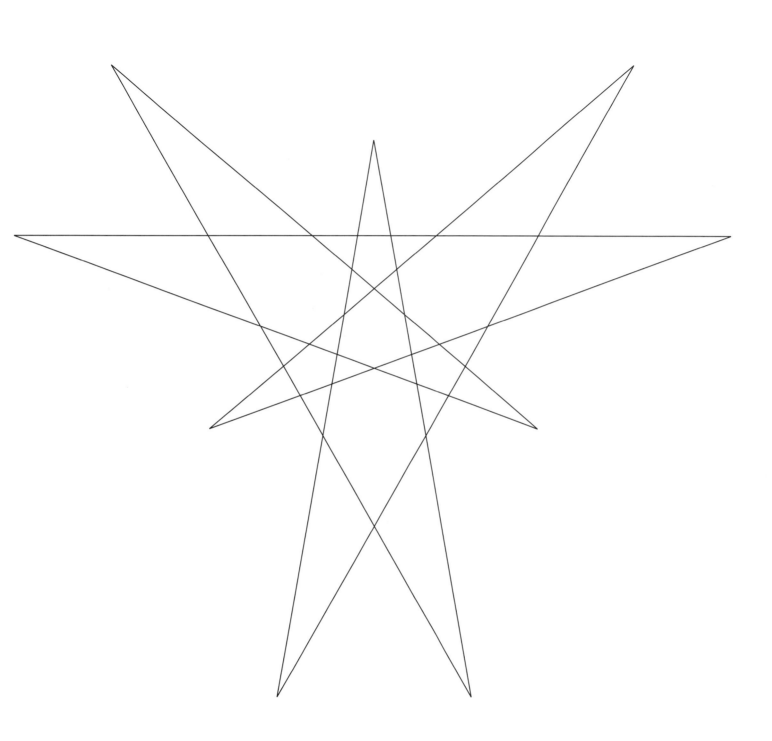

GRAPH THEORY

This type of graph uses nodes *joined by lines to represent a network of relationships*

BALABAN'S CAGE

Alexandru Balaban is an eminent Romanian chemist who likes to dabble in maths in his spare time. He discovered the first graph in which every node is connected to three others and in which the shortest loop has ten nodes. Known as the *Balaban 10-cage*, it is illustrated opposite.

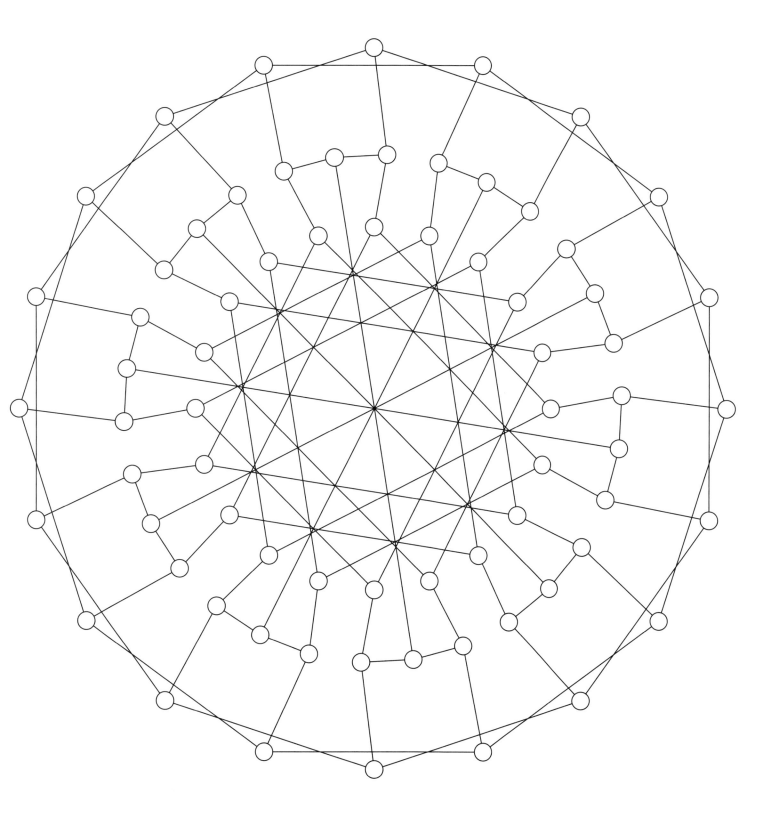

CALCULUS

The mathematics of measuring how fast things accumulate and how fast they change

NEWTON'S CUBIC

In 1666 Isaac Newton was holed up in his mother's farmhouse, avoiding the plague that was ravaging the cities and working on what he called his 'method of fluxions', a way to calculate the slope of curves. In one document called the *Tract of 1666* he investigated the properties of the curve $x^3 - abx + a^3 - cy^2 = 0$, where a, b and c are constants. The equation is known as a *cubic* because the largest power of x is a cube. Opposite is the family of curves produced by this equation when $a = 1$, $c = 4$ and b takes different values between -8 and 8. The method of fluxions later became known as *differentiation*, one of the basic ideas of calculus.

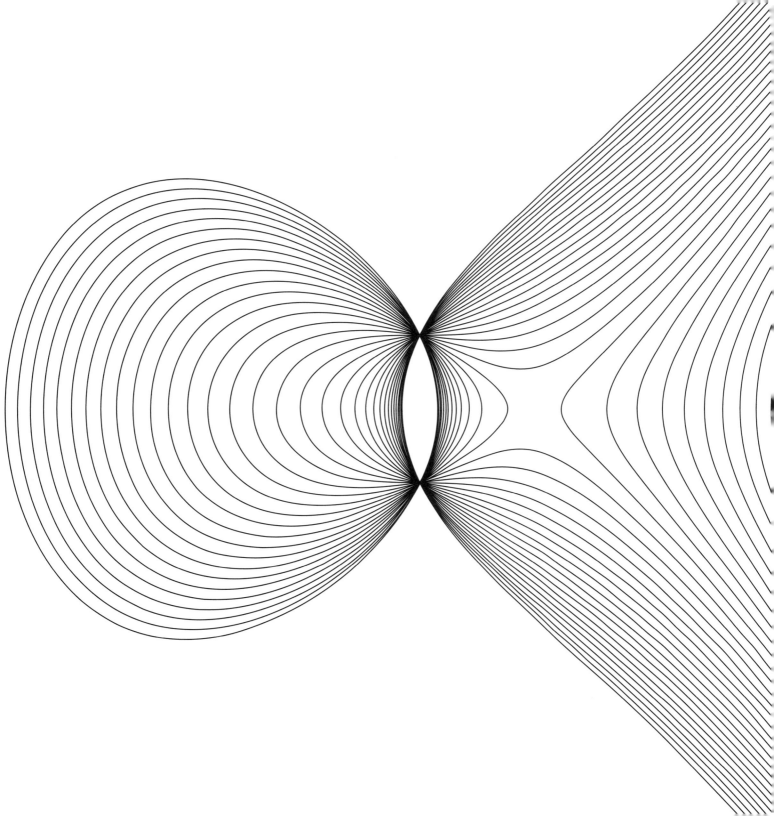

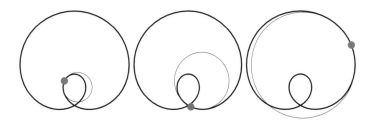

KISSING CIRCLES

The German scientist Gottfried Leibniz developed the ideas of calculus independently from Newton, but at the same time. In his work on curves he coined the term *circulus osculans* (Latin for 'kissing circle') to describe a circle that sits perfectly on a curve, meaning that the circle touches the curve at a single point where they share the same curvature. The image here is of many of these osculating – or kissing – circles on the bean-shaped curve above.

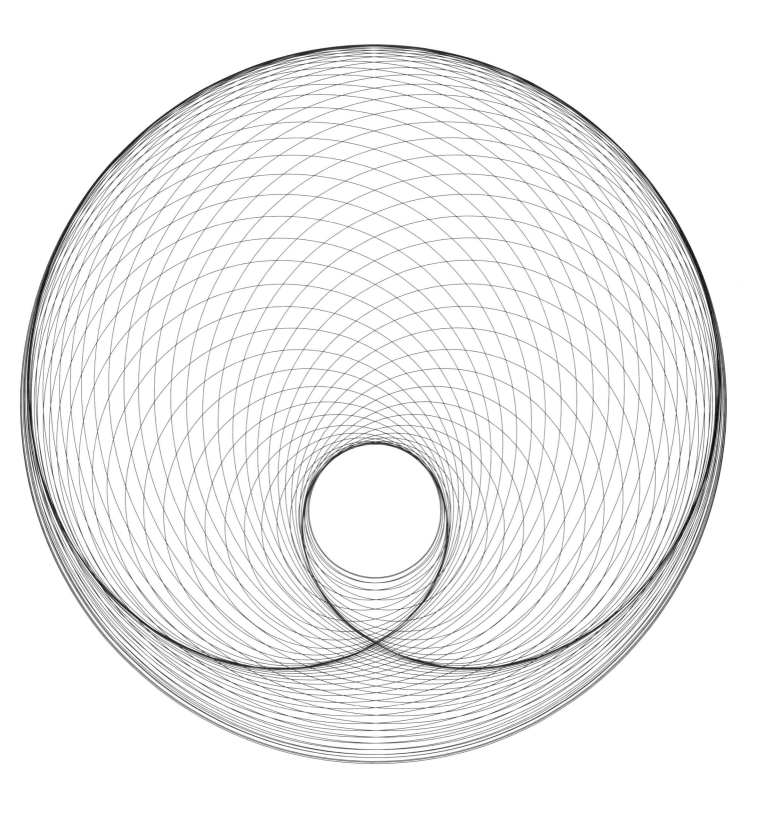

DYNAMICAL SYSTEMS

A dynamical system models motion or change over time

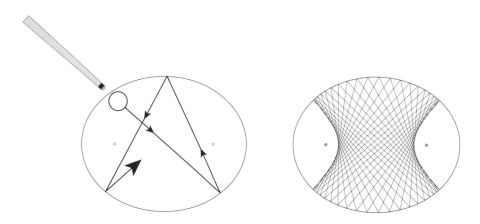

ELLIPTICAL POOL

Anyone for pool? Here, the shape of the table is an *ellipse*, and the two grey dots are its *focus points*. Since this is a mathematical pool table, we can ignore friction and imagine that, once in motion, our ball will continue to rebound forever. When you shoot a ball between the focus points (above), it will always rebound between them, and its complete path will trace out a *hyperbola* – two curves that are mirror images of each other.

When you shoot the ball *through* one focus, however, it will always rebound through the other focus. For this reason Alex decided to build an elliptical pool table with a pocket at one focus point and a black dot at the other. For more information on this exciting development in indoor sports, check out loop-the-game.com!

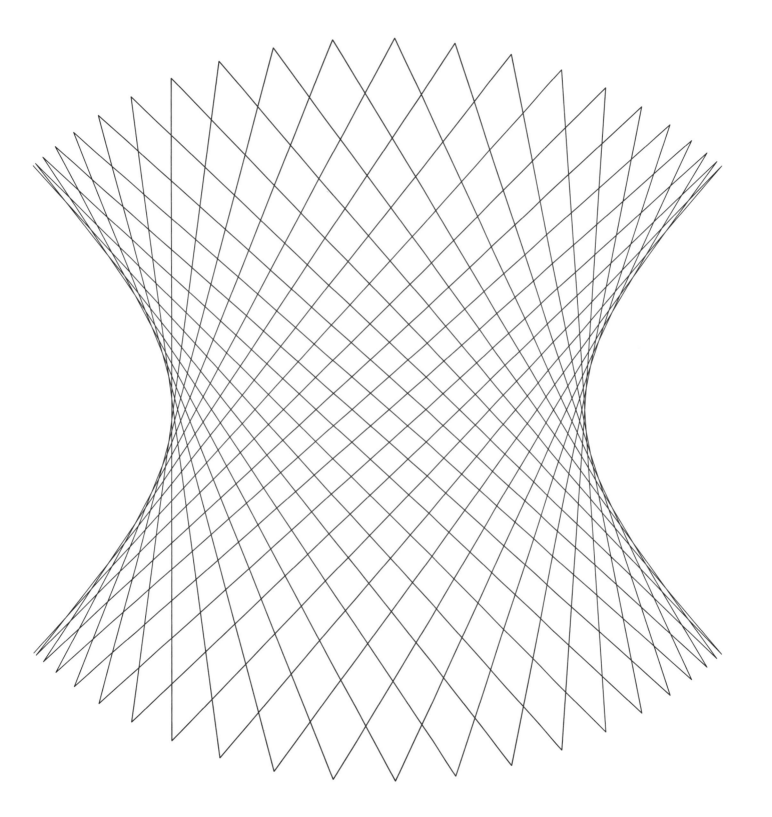

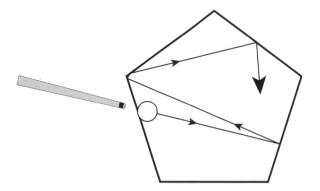

PENTAGONAL POOL

For our second round of mathematical pool, we have moved to a pentagonal table.

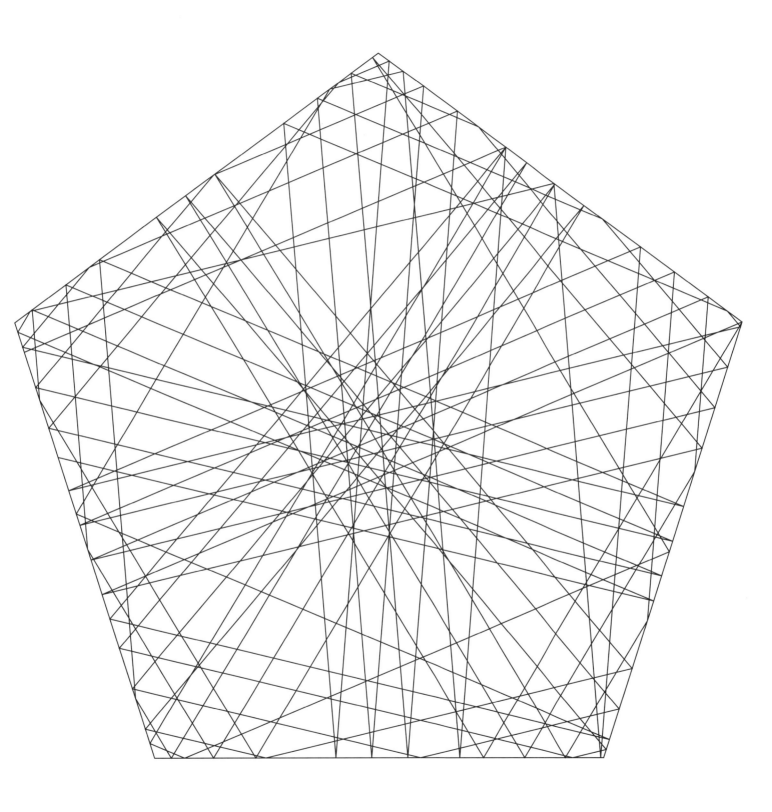

 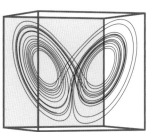 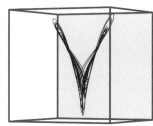 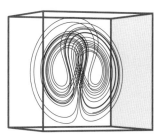

STRANGE ATTRACTOR

In 1963 Edward Lorenz developed a mathematical model for the convection currents that flow in the atmosphere as air warms up, rises, cools off and sinks down again. When his three equations – the *Lorenz equations* – take on certain values, the resulting behaviour becomes wild and chaotic, as in the butterfly image opposite. Other views of this *strange attractor* – the mathematical term for this type of pattern – are illustrated above.

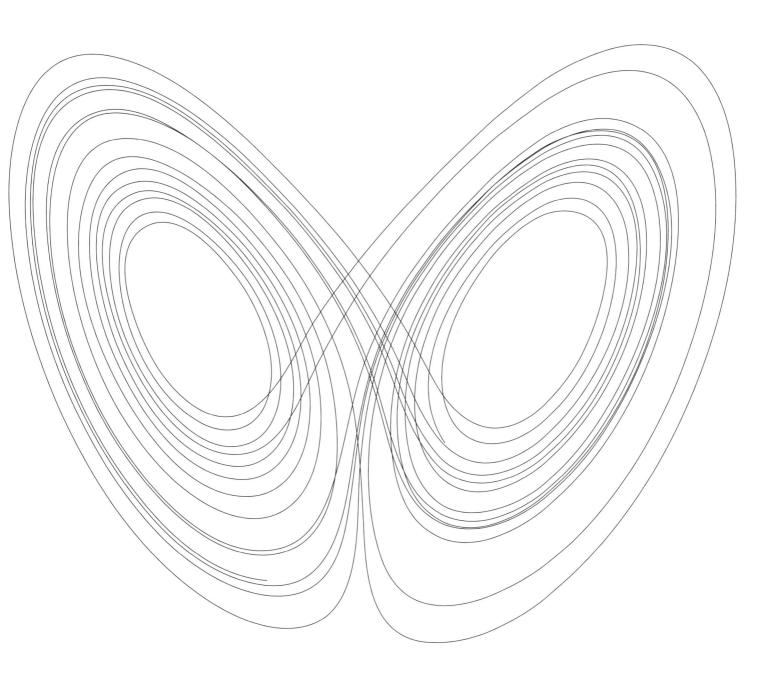

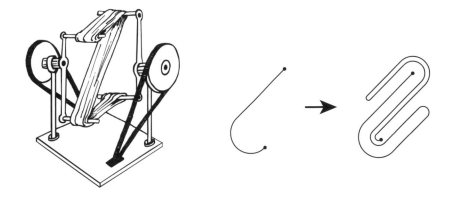

TASTY MATHS

Taffy candy is a traditional American sweet made by spinning a sugar-and-butter mix in a machine, like the one above, to create folds within folds within folds. The image opposite shows the mathematical structure of a cross section of the candy.

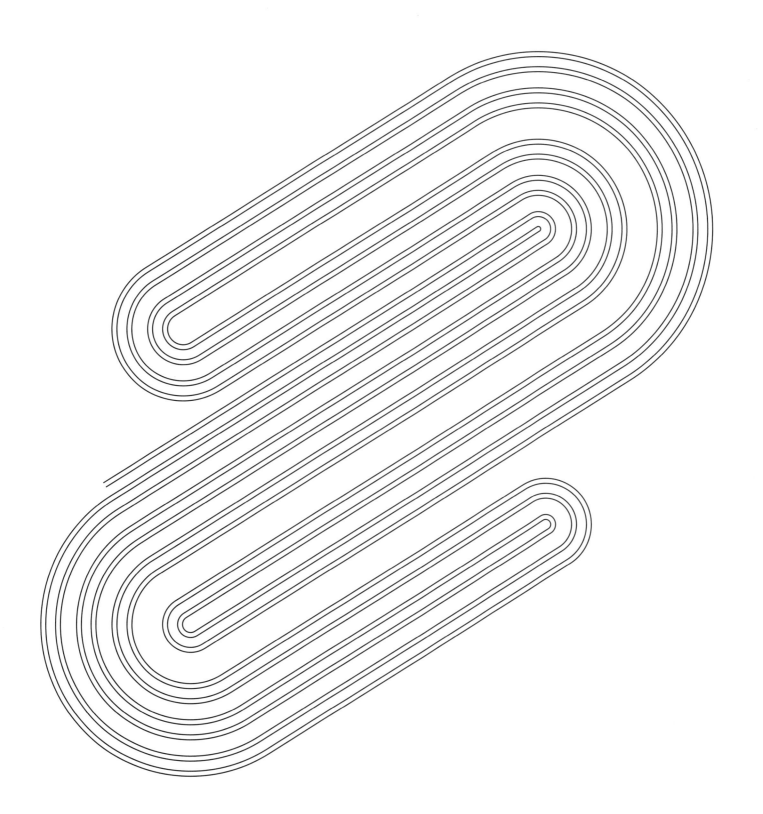

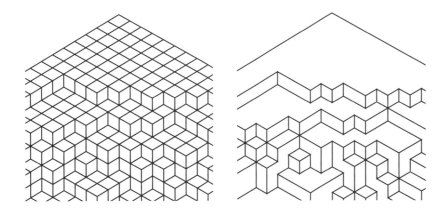

ARCTIC CIRCLE

Some dynamical systems model a change in structure. This one models the abrupt change — or *phase transition* — between solid and liquid when, say, ice melts. Here, a big hexagon is tiled with identical small rhombus-shaped tiles. Each tile can point in one of only three possible directions. There are a vast number of possible tile arrangements that will fill the hexagon, but almost all of them have large sections of tiles 'frozen' together in the same orientation near the edges — hence the name 'Arctic Circle'. In other words, the hexagonal shape forces the structure in the corners, but not in the centre, giving an abrupt transition between order and chaos. This phenomenon can be seen clearly opposite, where adjacent tiles lying in the same orientation are joined together.

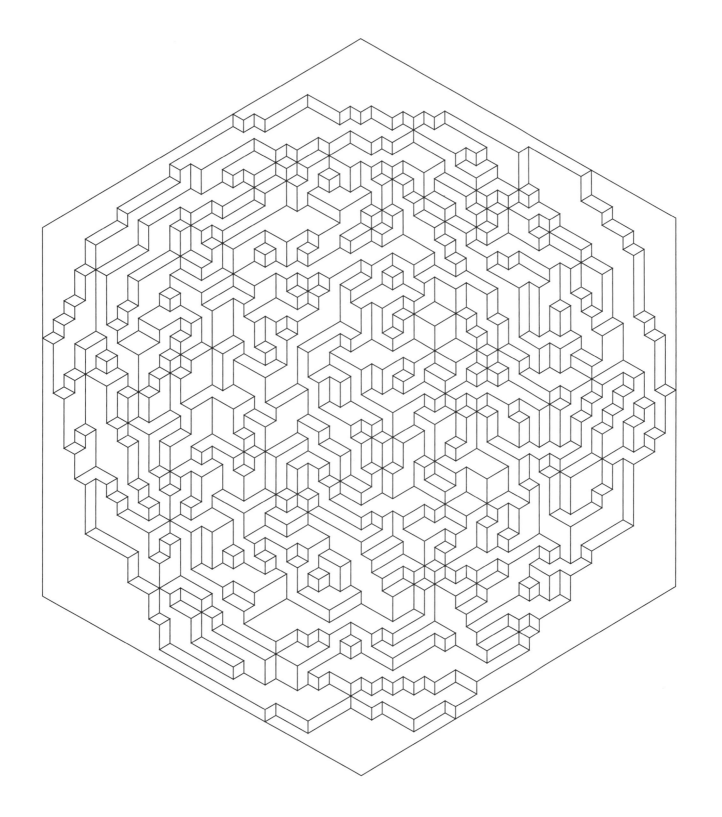

STATISTICAL PHYSICS

Using randomness to model physical phenomena, such as the behavior of gases

BROWNIAN TREES

In the beginning, there was a single tiny square in the centre of this image. Then a second square appeared at a random position at the edge of the page and started moving across it with *Brownian motion* — that is, in an erratic, random way. Eventually the moving square hit the central square and attached to it. Then a *third* square appeared at a random position at the edge of the page and followed a Brownian path until hitting, and attaching to, the central cluster of squares. This process is called *diffusion-limited aggregation*, and it continued until the final pattern emerged. Opposite, groups of squares have been drawn with a shared outline to represent stages of the pattern's growth. The distinctive-looking tendrils are called *Brownian trees*.

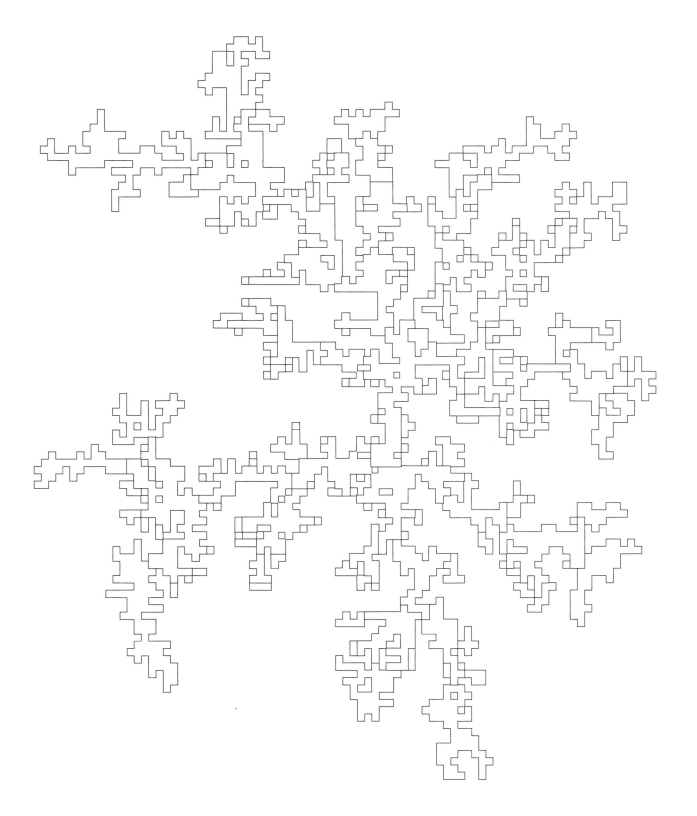

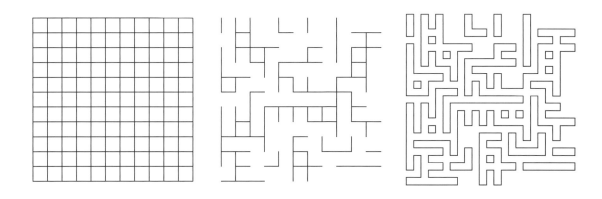

PERCOLATION

Start with a square grid (above left). Then delete 51 per cent of the edges at random, leaving a broken grid (centre). The final image (right) transforms the edges that remain into corridors by giving them *width* as well as length.

What's mathematically interesting here is that if you delete 50 per cent of the edges at random, then the chance of having an unbroken corridor from the top to the bottom of the final image is about 50/50. However, if you delete just over 50 per cent, that chance shrinks to almost zero, and if you delete just under 50 per cent, the chance rises to almost 100 per cent. We call this field *percolation theory* because it studies whether a liquid poured over the top of something porous will seep through to the bottom or not.

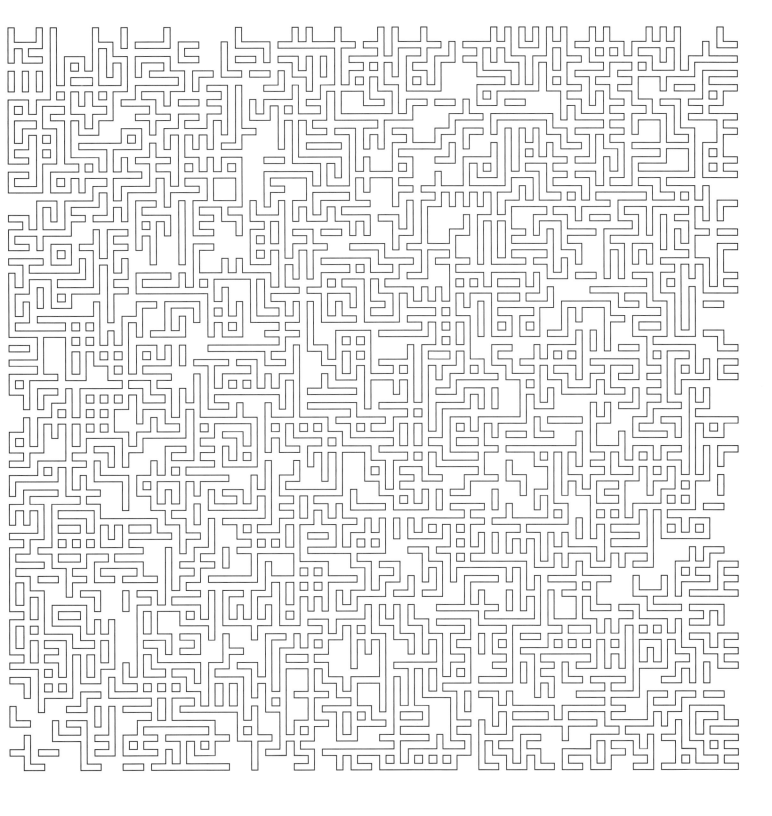

REACTION-DIFFUSION

A mathematical model of how a chemical reaction spreads

Computer science pioneer Alan Turing was the first to suggest that reaction-diffusion systems explain how fish and other animals get their stripes and spots.

NOODLE DOODLE

Chemicals react at different rates depending on how concentrated they are at any given point. Here, two chemicals are reacting and diffusing (spreading) evenly across the page. Each line surrounds a region dominated by one of the chemicals. This image is inspired by the work of Robert Munafo.

HODGE-SPLODGE
This reaction-diffusion system changes how the two chemicals react, going from the centre to the outside of the image.

ALGORITHMS

Proecesses that follow step-by-step rules

Computers use algorithms to process large data sets.

EDGE DETECTION

The edge-detection algorithm turns a photograph into a line drawing by measuring how quickly the colour changes between sets of pixels and drawing boundaries where it changes the fastest.

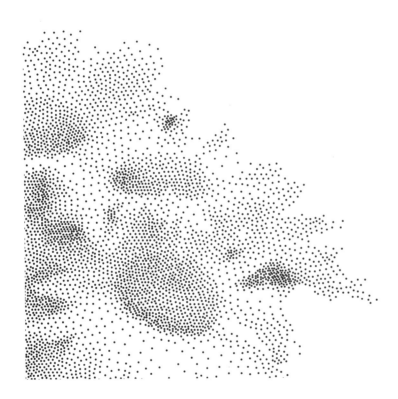

THE TRAVELLING SALESMAN

First, a 'stippling' algorithm turned the picture of flowers into 15,000 dots. Next, a different algorithm linked all the dots to create the shortest possible path that includes every dot only once. The problem of linking dots in this way is known as the *travelling salesman problem*, or TSP, since it was first posed about a salesman trying to make a round trip through several towns by the most efficient route.

PERLIN NOISE

The image above left is of white noise. In 1983, computer scientist Ken Perlin invented an algorithm that creates a more textured type of noise, illustrated above right, which is now often used in movie special effects to create natural-looking surfaces. The image opposite used a square of Perlin noise as its underlying structure.

CREATE

ISLAMIC GEOMETRY

Complex patterns that can be drawn with only a compass and straightedge

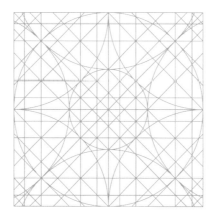

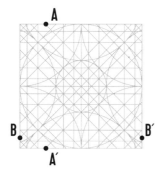

SQUARE TILE

Design a pattern for the tile template above, and then trace it sixteen times on the grid on the opposite page. (We've given you three examples for inspiration.) Be sure to plan your design so the lines will join up when you place the tiles next to each other – in other words, if a line touches the top of the square at A, make sure that a line also touches the bottom side of the square at A'. Likewise with B on the left and B' on the right.

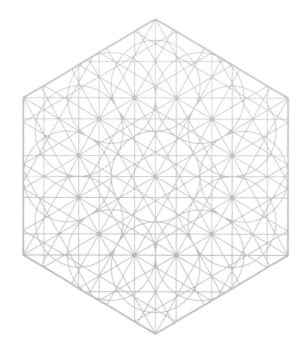

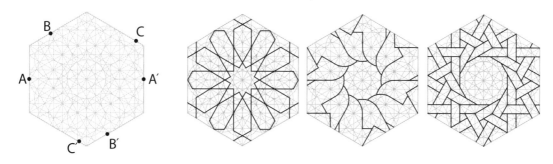

HEX TILE

Design a pattern for the tile template above just like you did for the previous image, and then fill in the grid on the opposite page. (Again, we've given you three examples.) This time, so that the lines will join up when you place the tiles next to each other, be sure to use matching points on parallel sides of the hexagon. That is, if a line touches A on the left side, make sure that a line also touches A′ on the right side. Likewise with B top left and B′ bottom right, and C top right and C′ bottom left.

GRAPH THEORY

Dots connected by lines

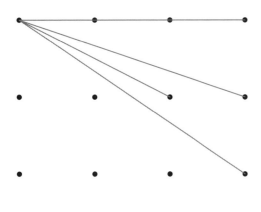

TWO'S COMPANY, SQUARE

Opposite, use a ruler to draw a line between every possible pair of dots.

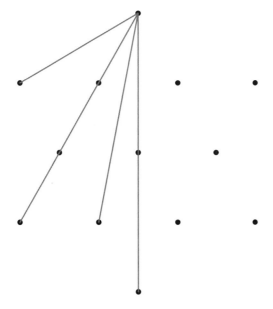

TWO'S COMPANY, HEXAGON

Opposite, use a ruler to draw a line between every possible pair of dots.

RANDOMNESS

Making pictures based on the flip of a coin

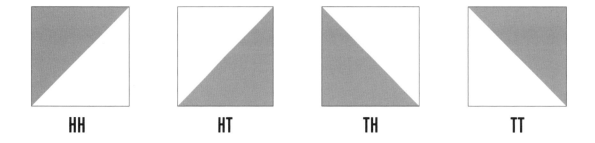

HH   HT   TH   TT

TRUCHET TILE

In 1704, Father Sébastien Truchet, a French priest and active scientist, described the tile above — a square split between two colours on the diagonal — and studied the patterns it could create. Here we'd like you to draw a Truchet tile in each of the squares on the grid opposite, at random. For each square, flip a coin twice and use the key above to determine the orientation of the tile. For example, if you get heads and then heads again, draw the tile labeled 'HH'.

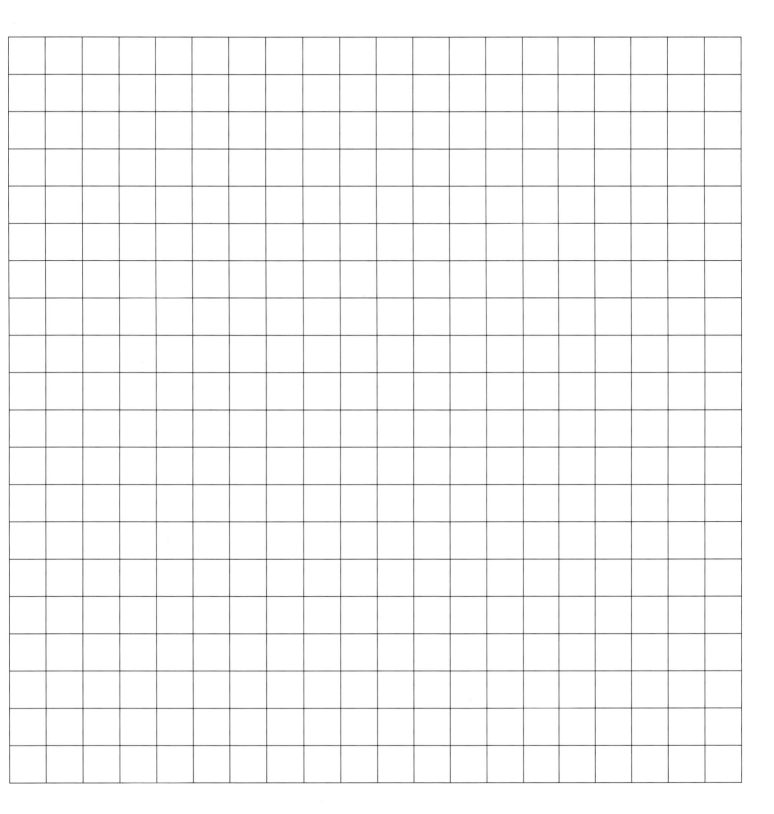

ALGORITHMS

An algorithm is a step-by-step rule

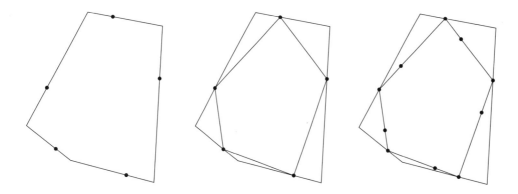

CONVEX SPIRAL

Working clockwise, mark a point roughly a third of the way along each side of the pentagon opposite, then use a ruler to join the points as above. Repeat with the new pentagon you have just created, and continue until the pentagons are as small as possible.

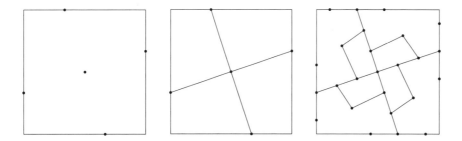

SQUARE SLICER

Working clockwise, mark a point roughly a third of the way along each side of the square opposite, then use a ruler to join each of them to the square's centre. This will create four *quadrilaterals*, as above. Repeat the process for each quadrilateral: That is, mark a point roughly a third of the way along each side, then join to roughly where the centre of the shape is. We've started this process above, joining two of the points to the centre of each quadrilateral. Join the other two points to get a total of 16 quadrilaterals. You can colour the sections now, or do it when you have carried out the process once, or even twice more, which would give you a stunning collage of 256 quadrilaterals. This process is analogous to algorithms used in computer animation.

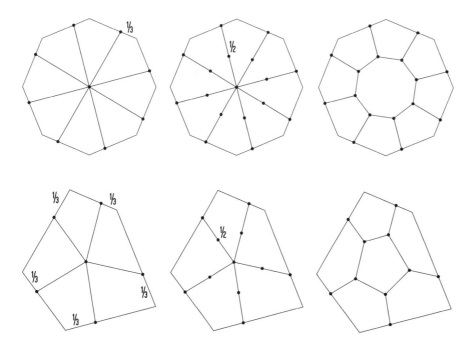

PENTAGON-A-GO-GO

We created the octagonal honeycomb opposite by starting with an octagon and following the steps above: Working anticlockwise, mark a point a third of the way along each edge. Link the points to the centre, then connect the midpoints of the new lines to make a smaller octagon, and erase inside it. Now it's your turn to repeat the process – opposite, follow the steps to turn the central octagon into a honeycomb, and then follow them once more with your *new* central octagon.

With the pentagons, follow the equivalent process, which is also shown above. Make dots a third of the way along each side, then join them to the centre, and connect the lines' midpoints to make a smaller pentagon. Try this twice on the outer eight pentagons (which we drew), and then once on the inner ring of pentagons (which you drew in your first honeycomb).

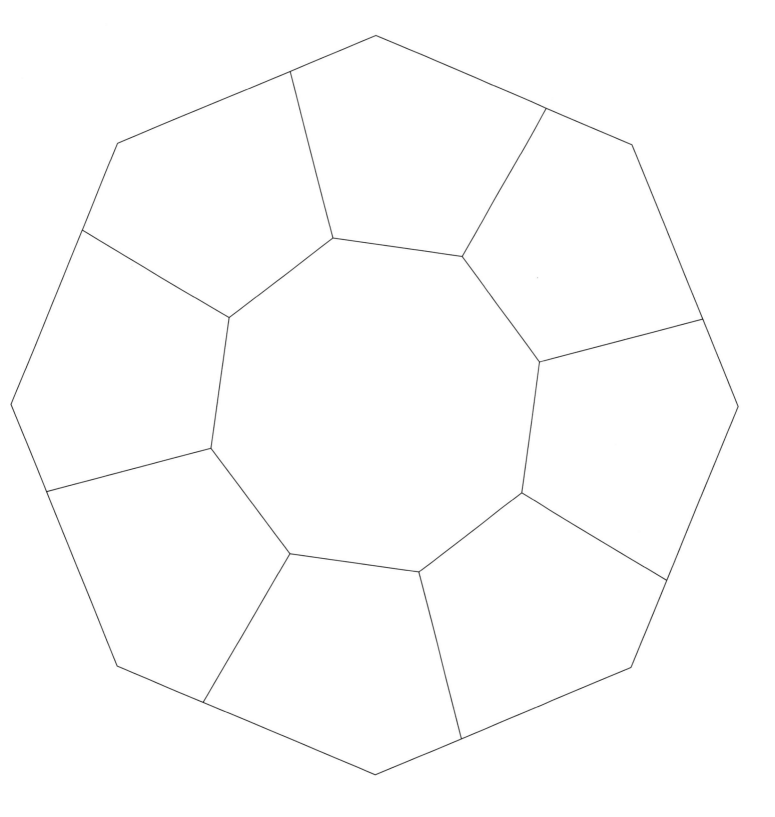

PASCAL'S TRIANGLE

A pyramid of numbers with amazing mathematical properties

PAINTING PASCAL

Inside Pascal's triangle, each number is the sum of the two numbers above it. For example, the number 6 on the fifth line has a 3 and a 3 above it. And the number 6 on the seventh line has a 1 and a 5. Pascal's triangle has many interesting properties, which you can discover by colouring it in. For example, try colouring all the odd numbers with one colour and see what you get. Or, if that sounds too simple, pick a colour for all the numbers divisible by 3, or all the numbers divisible by 4, or by 5.

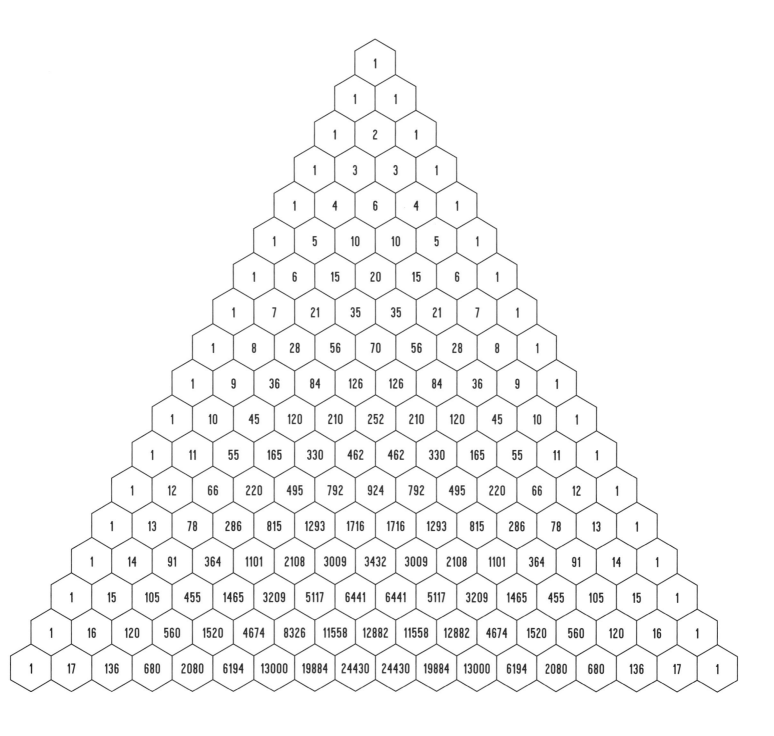

KNIGHT'S TOUR

A sequence of moves that takes the knight to every square on the chessboard exactly once

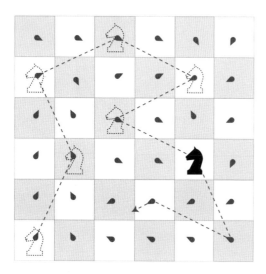

KNIGHT RIDER

In chess, the knight moves two squares in one direction and one square to the side. Interest in knight's tours dates from at least as far back as the ninth century, although serious mathematical research into the subject started with the great Swiss mathematician Leonhard Euler in the eighteenth century. When filled in, the image opposite will reveal a knight's tour of a 20 × 20 chessboard. The pointer on each square indicates the direction of the next move. Choose any square to start, and then 'join the dots' by following the direction of the pointers, as shown in the example above.

FURTHER JOURNEYS THROUGH THE WORLD OF MATHS

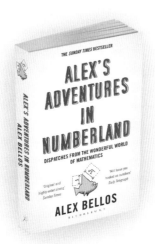

ALEX'S ADVENTURES IN NUMBERLAND

'Original and highly entertaining' **Sunday Times**

'A page turner about humanity's strange, never easy and, above all, never dull relationship with numbers' **New Scientist**

'Outstanding ... laced with humour, but at all times the star of the show is mathematics' Ian Stewart, **Prospect**

ALEX THROUGH THE LOOKING GLASS

'Immediate, relevant and fun' **Daily Telegraph**

'To read *Alex Through the Looking Glass* is to have one's mind quietly but continually blown' **Sunday Express**

'Alex Bellos brings the quirks and eccentricities of numbers wonderfully to life' Simon Singh, **Observer**

B L O O M S B U R Y